Cement Industry - Optimization, Characterization and Sustainable Application

Edited by Hosam El-Din Mostafa Saleh

Published in London, United Kingdom

IntechOpen

Supporting open minds since 2005

Cement Industry - Optimization, Characterization and Sustainable Application
http://dx.doi.org/10.5772/intechopen.87890
Edited by Hosam El-Din Mostafa Saleh

Contributors
Hilal El-Hassan, M.Jemimah Carmichael, G. Prince Arulraj, Shankar Sabavath, Chakravarthi Sarella, Raj Kumar Galipelli, Muhammad Imran Khan, Muslich Hartadi Sutanto, Madzlan Bin Napiah, Salah E. Zoroob, Michael Mikheenkov, Oleg Sheshukov, Gude Reddy Babu, Nelluru Venkata Ramana, Bhumireddy Madhusudana Reddy, Pala Gireesh Kumar, Ahed Habib, Maan Habib, Jaime Andres Perez Taborda, Jorge Hernan Quintero Orozco, Daniel Andres Triana Camacho, Michal Bacuvcik, Pavel Martauz, Ivan Janotka, Branislav Cvopa, Oluwafemi Fadayini, Adekunle Adedapo Obisanya, Gloria Ajiboye, Clement Madu, Tajudeen Ipaye, Taiwo Rabiu, Joseph Taiwo Akintola, Shola John Ajayi, Taiwo Oshin, Adekunle Obisanya, Nkechi Alexandra Kingsley, Hosam M. M. Saleh, Abeer M. El-Sayed, Abeer A. Faheim, Aida A. Salman

First published in London, United Kingdom, 2021 by IntechOpen
IntechOpen is the global imprint of INTECHOPEN LIMITED, registered in England and Wales, registration number: 11086078, 5 Princes Gate Court, London, SW7 2QJ, United Kingdom
Printed in Croatia

British Library Cataloguing-in-Publication Data
A catalogue record for this book is available from the British Library

Additional hard and PDF copies can be obtained from orders@intechopen.com

Cement Industry - Optimization, Characterization and Sustainable Application
Edited by Hosam El-Din Mostafa Saleh
p. cm.
Print ISBN 978-1-83962-314-1
Online ISBN 978-1-83962-315-8
eBook (PDF) ISBN 978-1-83962-328-8

For EU product safety concerns:
IN TECH d.o.o., Prolaz Marije Krucifikse Kozulić 3, 51000 Rijeka, Croatia, info@intechopen.com or visit our website at intechopen.com.

Meet the editor

Hosam Saleh is a Professor of Radioactive Waste Management in the Radioisotope Department, Atomic Energy Authority, Egypt. He obtained an MSc and Ph.D. in Physical Chemistry from Cairo University. Dr. Saleh has more than 23 years of experience in hazardous waste management with an emphasis on treatment and developing new matrixes for immobilizing these wastes. He is also interested in studying innovative economic and environmentally friendly techniques for the management of hazardous and radioactive wastes. He has authored many peer-reviewed scientific papers and chapters and served as an editor for several books from international publishers. He has been selected among the top 2 percent of scientists in the world, according to the Stanford University report for 2020.

Contents

Preface

Extensive studies have been reported to improve various unpleasant characteristics of cement-based composites, aiming at efficient utilization of these novel materials, especially in the case of solidification/stabilization of hazardous waste. This book shows how cement, one of the key materials for sustainable economic and environmental development in different global regions, can be purposefully modified and implemented. It shows that smart and sustainable methods are needed to make such innovative processes economical while at the same time maximizing environmental and human health benefits.

The authors have summarized their experience and present advances in relevant fields related to assessing the strategies of sustainable applications of the cement industry.

The book contains twelve chapters, organized in four sections that cover important research aspects of the cement industry related to its optimization, characterization, and sustainable application. The first section includes the introductory chapter prepared by the editor to present a brief background on the cement industry.

Section 2 "Characterizations of New Cement Compositions" includes the following five chapters: "Compressive Strength of Concrete with Nano Cement" by Jemimah Carmichael Milton and Prince Arulraj Gnanaraj;

"The Resistance of New Kind of High-Strength Cement after 5 Years Exposure to Sulfate Solution" by Michal Bačuvčík, Pavel Martauz, Ivan Janotka and Branislav Cvopa

"Impact of Nanosilica in Ordinary Portland Cement over Its Durability and Properties" by Gude Reddy Babu et al.

"Simulation and Optimization of an Integrated Process Flow Sheet for Cement Production" by Oluwafemi M. Fadayini et al.

"Peculiarities of Portland Cement Clinker Synthesis in the Presence of a Significant Amount of SO_3 in a Raw Mix" by Oleg Sheshukov and Michael Mikheenkov.

Section 3 "Comparative Parameters of Cement Production" includes the following three chapters: "Energy and Economic Comparison of Different Fuels in Cement Production" by Oluwafemi M. Fadayini et al.

"Accelerated Carbonation Curing as a Means of Reducing Carbon Dioxide Emissions" by Hilal El-Hassan

"Cement-Based Piezoelectricity Application: A Theoretical Approach" by Daniel A. Triana-Camacho, Jorge H. Quintero-Orozco and Jaime A. Perez-Taborda.

Section 4 "Sustainable Applications of Cement" includes the following three chapters:

"Sustainable Recycling of Marble Dust as Cement Replacement in Concrete: Advances and Recent Trends" by Ahed Habib and Maan Habib.

"Applications of Cement in Pavement Engineering" by Sarella Chakravarthi, Galipelli Raj Kumar and Sabavath Shankar

"Cementitious Grouts Containing Irradiated Waste Polyethylene Terephthalate" by Muhammad Imran Khan, Muslich Hartadi Sutanto, Madzlan Bin Napiah and Salah E. Zoorob

The editor wishes to thank all the participants in this book for their valuable contributions and Ms. Sara Debeuc, Author Service Manager at IntechOpen, for her continuous assistance in finalizing this work.

Hosam El-Din Mostafa Saleh
Egyptian Atomic Energy Authority,
Cairo, Egypt

Section 1

Introduction

Chapter 1

Introductory Chapter: Cement Industry

Abeer M. El-Sayed, Abeer A. Faheim, Aida A. Salman and Hosam M. Saleh

1. Introduction

Cement is a capital-intensive, energy-consuming and critical sector for the construction of nation-wide infrastructure. The international cement industry, while constituting a limited share of the world's output has been rising at an increasing pace compared to the local demand in recent years. Attempts to protect the environment in developing countries, particularly Europe have forced cement manufacturing plants to migrate to countries with less strict environmental regulations. Along with consistently rising real prices, this has provided a trend for economic performance and environmental enforcement [1].

It is worth noting that cement is known to be one of the most important construction materials in the world. It is primarily used in the manufacture of concrete. Concrete is a combination of inert mineral aggregates such as sand, gravel, crushed stones and cement. Consumption and production of cement are directly connected to the building sector and thus to the general economic activity. Cement is one of the most developed goods in the world, due to its importance as a building material and the geographical availability of the main raw materials, i.e. limestone, cement is manufactured in almost all countries. The widespread development is also due to the comparatively low price and high density of cement, which, due to the relatively high costs, decreases ground transport. Export trade (excluding border-based plants) is typically limited relative to global production.

Cement-based materials, such as concrete and mortars, are used in very significant amounts. For example, concrete production amounted to more than 10 billion tonnes in 2009. Cement plays an important role in terms of economic and social importance as it is necessary to develop and enhance infrastructure. This sector, on the other hand, is also a strong polluter. Cement processing emits 5–6% of the carbon dioxide emitted by human activity, accounting for around 4% of global warming. It may emit vast quantities of chronic chemical contaminants, such as dioxins and heavy metals and particulate matter. Energy use is also important. Cement production accounts for about 0.6% of all electricity generated in the United States. In the other hand, the chemistry driving the manufacture of cement and its applications can be very beneficial in solving these environmental concerns.

Cement manufacturing is an extremely energy-intensive method of processing. The energy consumption is measured at around 2% of global primary energy consumption, or approximately 5% of total manufacturing energy consumption [2], regarding to the prevalent use of carbon-intensive fuels, e.g. coal, in the manufacture of clinkers. In addition to energy consumption, the clinker process also releases CO_2 as a result of the calcination process. Ecofys Energy and Climate and Berkeley National Laboratory therefore carried out an appraisal for the IEA Greenhouse

Gas R&D Program on the role of the cement industry in the development of CO_2 and the options for lowering carbon dioxide emissions. This discuss the historical development and global distribution of cement production [3].

Moreover, the cement industry needs raw materials, fuel and chemical additives, and these activities generate emissions which have a negative effect on the quality of the atmosphere. The gas emissions emitted are CO_2, CH_4, NOx, SOx, N_2O and particulate matter.

These emissions have an effect on the rise in global warming and the decrease in atmospheric air quality, which has an impact on human health and the atmosphere [4].

However, cement is the second primary source of anthropogenic pollution, source for about 7% of global CO_2 emissions. The technology for carbon dioxide capture and storage (CCS) is considered by the International Energy Agency (IEA) to be a crucial technology capable of lowering CO_2 emissions in the cement sector by 56% by 2050. CO_2 capture technologies for the cement production process and analyses economic and financial problems relevant to carbon dioxide capture in the cement production has an important trend for study [5].

The overall CO_2 emissions from cement manufacturing, including process and energy-related emissions has a significant interest. Actually, much of the relevant evidence only covers process pollution. CO_2 pollution control solutions for the cement industry are also discussed. In 1994, the projected gross carbon emissions from cement manufacturing is 307 million metric tonnes of carbon (MtC), 160 MtC from process carbon emissions and 147 MtC from electricity usage. Overall, the top 10 cement-producing countries accounted for 63% of the total carbon emissions from cement manufacturing in 1994. The estimated strength of carbon dioxide emissions from global cement output is 222 kg of C/t of cement. Emissions reduction solutions include enhancing energy quality, new methods, transitioning to low-carbon oil, using waste oils, the use of additives in cement processing, and gradually eliminating substitute cements and CO_2 from flue gas in clinker kilns [6].

Contamination of the atmosphere in the area of cement factories, e.g. some cement emissions around it, it may be claimed that CaO percentages were found to be higher (37.7%) particularly in surface soil samples taken near the cement factory. Based on the geo-accumulation index, soils in the study area could be graded as moderately to highly contaminate with (As, Cd, Pb and Ni) and highly contaminated with Cr, whereas soils in the study region were moderately polluted with Zn. On the other hand, the soils of the sample region are considerably polluted with As, Cd and Cu ($5 > EF > 20$) on the basis of the Enrichment Factor (EF). The most hazardous areas are clustered within 0 to 2 km of the cement plant [7].

As the health history of factory employees and certain inhabitants of nearby areas indicates a high incidence of respiratory and skin infections. Regulation of the regulations on pollution enforcement and the establishment of a buffer zone around the cement factory can protect the atmosphere and public health [8].

Egypt increased cement production from 4 million tonnes in 1975 to 46 million tonnes in 2009 and now accounts for about 1.5 percent of global cement supply. Dust emissions account for around 6% of PM10 in Greater Cairo, hitting as much as 30% in areas near cement plants. New regulatory requirements, due to be approved in 2010, would-the emissions of dust from 300 to 100 mg/m^3 for existing plants and from 100 to 50 mg/m^3 for new plants. Online tracking of the 72 main stacks in the 16 cement plants by the Egyptian Environmental Affairs Agency (EEAA) offers real-time details on the emissions of carbon. New plants are 98% compliant and older plants are 92% compliant with pollution standards. No manual monitoring of SO_X and NO_X pollution is performed. Cleaner development and pollution control prospects for the cement sector include: i) the use of alternative fuels in cement kilns; ii) the reduction of NO_X; iii) the removal of dust emissions; iv) the use of silica waste to manufacture new cement products; v) the reuse of bypass dust; and vi) the disposal of radioactive waste [9].

As far as processing is concerned, there are many alternate products that can be used to mitigate carbon dioxide emissions and limit energy consumption, such as calcium sulfoaluminate and b-Ca_2SiO_4-rich cements. The use of residues from other manufacturing industries will also increase the sustainability of the cement industry. Under suitable conditions, waste materials such as tires, fuels, urban solid waste and solvents can be used as additional fuel in cement plants. Concrete can be used to encapsulate discarded products such as rubber, plastics and glasses. In this manner, certain aspects of the cement industry related to environmental science are explored. Other problems, such as economic considerations, the chemistry of cement manufacturing and its properties, are also addressed. Special attention is paid to the role that cement chemistry can play in terms of sustainability. The most important elements, such as the use of substitute products, are outlined; fresh opportunities as well as the recycling of products. It is also argued that the role of research and development required to boost the sustainability of cement is a significant feature [10].

Author details

Abeer M. El-Sayed[1], Abeer A. Faheim[1], Aida A. Salman[1] and Hosam M. Saleh[2*]

1 Chemistry Department, Faculty of Science, Al Azhar University, Egypt

2 Radioisotope Department, Nuclear Research Center, Atomic Energy Authority, Giza, Egypt

*Address all correspondence to: hosam.saleh@eaea.org.eg; hosamsaleh70@yahoo.com

References

[1] T. Selim and A. Salem, "Global cement industry: Competitive and institutional dimensions," 2010.

[2] N. Martin, M. D. Levine, L. Price, and E. Worrell, "Efficient use of energy utilizing high technology: An assessment of energy use in industry and buildings," *London World Energy Counc.*, 1995.

[3] C. A. Hendriks, E. Worrell, D. De Jager, K. Blok, and P. Riemer, "Emission reduction of greenhouse gases from the cement industry," in *Proceedings of the fourth international conference on greenhouse gas control technologies*, 1998, pp. 939-944.

[4] C. Chen, G. Habert, Y. Bouzidi, and A. Jullien, "Environmental impact of cement production: detail of the different processes and cement plant variability evaluation," *J. Clean. Prod.*, vol. 18, no. 5, pp. 478-485, 2010.

[5] J. Li, P. Tharakan, D. Macdonald, and X. Liang, "Technological, economic and financial prospects of carbon dioxide capture in the cement industry," *Energy Policy*, vol. 61, pp. 1377-1387, 2013.

[6] E. Worrell, L. Price, N. Martin, C. Hendriks, and L. O. Meida, "Carbon dioxide emissions from the global cement industry," *Annu. Rev. energy Environ.*, vol. 26, no. 1, pp. 303-329, 2001.

[7] A. M. Al-Omran, S. E. El-Maghraby, E. A. Nadeem, A. M. El-Eter, and S. M. I. Al-Qahtani, "Impact of cement dust on some soil properties around the cement factory in Al-Hasa Oasis, Saudi Arabia," *Am. J Agric Env. Sci*, vol. 11, no. 6, pp. 840-846, 2011.

[8] O. Oguntoke, A. E. Awanu, and H. J. Annegarn, "Impact of cement factory operations on air quality and human health in Ewekoro Local Government Area, South-Western Nigeria," *Int. J. Environ. Stud.*, vol. 69, no. 6, pp. 934-945, 2012.

[9] Y. Askar, P. Jago, M. M. Mourad, and D. Huisingh, "The cement industry in Egypt: Challenges and innovative cleaner production solutions," in *Knowledge Collaboration & Learning for Sustainable Innovation: 14th European Roundtable on Sustainable Consumption and Production (ERSCP) conference and the 6th Environmental Management for Sustainable Universities (EMSU) conference, Delft, The Netherland*, 2010.

[10] F. A. Rodrigues and I. Joekes, "Cement industry: sustainability, challenges and perspectives," *Environ. Chem. Lett.*, vol. 9, no. 2, pp. 151-166, 2011.

Section 2

Characterizations of New Cement Compositions

Chapter 2

Compressive Strength of Concrete with Nano Cement

Jemimah Carmichael Milton and Prince Arulraj Gnanaraj

Abstract

Nano technology plays a very vital role in all the areas of research. The incorporation of nano materials in concrete offers many advantages and improves the workability, the strength and durability properties of concrete. In this study an attempt has been made to carry out an experimental investigation on concrete in which cement was replaced with nano sized cement. Ordinary Portland cement of 53 grade was ground in a ball grinding mill to produce nano cement. The characterization of nano cement was studied using Scanning Electron Microscope (SEM), Brunauer Emmett–Teller (BET), Energy Dispersive X ray microanalysis (EDAX) and Fourier Transform Infrared Spectroscopy (FTIR). From the characterization studies, it was confirmed that particles were converted to nano size, the specific surface area increased and the chemical composition remained almost the same. The properties of cement paste with and without nano cement were found. For the experimental study, cement was replaced with 10%, 20%, 30%, 40% and 50% of nano cement. Cement mortar of ratio 1:3 and concrete of grades M20, M30, M40 and M50 were used. Compressive strength of cement mortar and concrete with different percentages of nano cement was found. The cement mortar was also subjected to micro structural study. It was found that the strength increased even up to the replacement level of 50%. Further increase in the replacement is not possible since the addition of nano cement reduces the initial and final setting time of cement paste. At 50% replacement level, the initial setting time got reduced to 30 minutes which the least permitted value as per IS 12269: 2013. The increase in strength was due to the fact that nano cement acts not only as a filler material but also the reactivity increased due to the higher specific surface area. The SEM image shows the formation of additional C-S-H gel. The percentage increase in compressive strength was found to increase up to 32%. The workability of concrete with nano cement was found to be significantly more than that of the normal cement concrete.

Keywords: compressive strength, nano cement, normal cement concrete, scanning electron microscope (SEM), energy dispersive X ray microanalysis (EDAX), Fourier transform infrared spectroscopy (FTIR)

1. Introduction

Nano technology is a new emerging area in field of engineering. Development of nanotechnology in the field of material science and evolution of advanced instrumentation have paved way for application of nanotechnology in the construction field. Incorporation of nano sized particles in cement composites makes a significant change in structural and nonstructural properties of cement paste, mortar and concrete. The particles converted from micron size to nano size results in more specific surface area. The increase in surface area leads to changes in morphology, increase in the chemical reactivity, structural modification of cement hydrates and enhancement of the properties of concrete. Nano particles are produced by two approaches. In "top down" approach, larger particles are reduced to smaller particles without altering the original properties and in "bottom down" approach very small nanoscopic molecules and atoms combine together to form bigger structures wherein the particles properties can be altered. The nano scale particles can result in vividly improved properties from conventional grain size materials of the same composition. Nano materials show unique physical and chemical properties that can lead to the development of more effective materials than the ones which are currently available. The use of nano materials in concrete, results in stronger and more durable concrete with desired stress-strain behaviour.

The structure of nano materials can be studied using the various sophisticated non-destructive techniques. The scaled down particles are to be checked for their size and some of the equipment available to determine particle size are scanning electron microscope, atomic force microscope and transmission electron microscope. Many techniques like Energy dispersive X-ray analysis, X-ray diffraction, X-ray absorption spectroscopy, Fourier transform infrared spectroscopy, Nuclear magnetic resonance *spectroscopy,* Thermal gravimetric analysis, Low-energy ion scattering spectroscopy, UV-V's spectroscopy, Photoluminescence spectroscopy, Dynamic light scattering are available for the surface chemical analysis and characterization of nano materials.

Nowadays application of nanotechnology can be widely seen in medical, car manufacturing, pharmaceutical, chemical and other industries. Nano particles are used for the manufacturing of medicines, bio medical instruments, paints, coatings, glass, plastics and rubber. In the construction field, nano titanium oxide, nano silica, nano aluminum oxide, nano zirconium oxide, carbon nano tubes, carbon nano fibers and nano fly ash are commonly used nano materials. According to Konstantin Sobolev [1], nano particles improve the ductility, thermal resistance and hence can be used in refractory concrete. Hanus [2] reported that nano particles produce anti-microbial surfaces and can be adopted in hospital buildings. It was also suggested that nano particles can be used in pavement as it possess high thermal resistance and abrasion property. According to Gann [3], shells and bones contain crystals of calcium and can be used in nano form to arrest crack and to dissipate energy.

Nano particles are added to concrete to improve its material properties. Perumalsamy Balaguru and Ken Chong [4] expressed that particle size upto 500 nm can be used in concrete whereas Surinder Mann [5], Florence Sanchez and Konstanin Sobolev [6] Bhuvaneshwari et al., [7], reported that the size of nano particles used in concrete has to be restricted to 200 nm. Hui li et al., [8], Maile Aiu and Huang [9] studied the effect of nano materials on the compressive strength of cement mortar. Tao Ji [10] and Ali Nazari and Shadi Riahi [11] studied the permeability and microstructure of concrete containing nano particles. Thomas.K.paul et al., [12] reported that nano materials can be produced using different techniques and gave an outline of producing nano fly ash. Research proves that there will be considerable changes in the chemical reactivity and mechanical properties, when the particles are converted to nano size. When the size is reduced, more atoms will be found at the surface of particle which significantly imparts a change in the morphology and energy at the surface. The changes in the chemical reactivity will improve the catalytic ability in paints and pigments which impart self-cleaning and self-healing properties. Nano titanium oxide is used as a self-cleaning material in glass and as anti-corrosive element in steel. Carbon nano tubes (both single walled and multi walled) and carbon nano fibers are used in concrete to prevent cracks and to improve ductility. Nano ferric oxide due to its super para magnetic property can improve the strength property of concrete. Nano aluminum oxide is used to resist abrasion of concrete in pavements. Nano fly ash acts as a promoter to enhance the pozzolanic property. Nano silica with different quality and properties are used as an additive material in concrete. It can be seen from literature that addition of nano particles significantly improves the strength and durability of concrete. In common, nano particles are used to enhance mechanical and durability properties of concrete and to develop sustainable concrete materials and structures.

2. Review of literature

The review of literature on the behaviour of mortar and concrete with nano materials is reported to understand the behaviour of cement mortar and concrete with nano particles, According to Hanus and Andrew T. Harris [13], Silvestre et al. [14] and Elzbieta Horszczaruk [15], nano silica, nano titanium oxide, nano zirconium oxide, carbon nano tubes and carbon nano fibers are commonly used materials for making nano concrete. These nano materials are not cost effective and also not available in abundance compared to the supplementary cementitious materials. Balaguru Perumalsamy and Chong Ken [4] reported that nano silica in colloidal state was more efficient than the micron sized silica in improving the durability of concrete. It was also reported that nano cement and nano carbon tubes can enhance the properties of concrete compared to nano carbon fibers. Konstantain Sobolev and Miguel Ferrada Gutierrez [16], Surinder Mann [5], Zhi Ge and Zhili Gao [17] reported that introduction of nanotechnology may bring major changes in the field of concrete technology. The concept of nanotechnology can be applied to concrete, steel and glass to produce new products with new properties. It was highlighted that the structure and behaviour of concrete should be understood at micro and nano scale. Bhuvaneshwari et al. [7] reported that efforts were taken to evolve new nano material towards a green and sustainable solution in the area of cement based materials and their composites for the construction applications. It was reported that the study of cement based material at nano level will result in a new generation construction materials with enhanced strength and durability properties. Sanchez Florence and Sobolev Konstantin [6] reported that the measurement and characterization of nano structure of cement and concrete materials is nano science and use of nano materials in cement and concrete composite is nano engineering. The nano structure study of nano materials was done by atomic force microscope, nano indentation technique, nuclear techniques, neutron and X ray scattering technique. Experimental study and micro structural study on nano materials is essential to study the effect of nano particles in concrete. They reported that mechanical properties of concrete can be enhanced with incorporation of nano particles in concrete. Laila Raki et al. [18] reported that nano-sized particles modify and improve the durability of concrete.

Jafarbeglou et al. [19] reviewed the current state of nano technology in enhancing the performance of concrete by producing new sustainable advanced cement based composites. The advance instruments like Atomic Force Microscope (AFM), Transmission Electron Microscope (TEM) were used to understand the role of nano particles and to predict the life of concrete with nano materials. It was discussed that the uniform proper dispersion and compatibility of nano materials should be taken care while incorporating the nano sized materials in concrete. Guillermo Bastos et al. [20] stated that it is necessary to have a unique synthesis method to produce nano materials in a large scale Standards should be adopted to mix efficiently nano particles in cement composites. High quality standards in production, common codes and identical terminology are needed to transfer the knowledge of research findings to global market. Implementation of concept of nano technology in construction field is difficult due to the cost involved in synthesis and dispersion of nano materials. Muhd Norhasri Muhd Sidek et al. [21] reported that particle size less than 500 nm can be used in concrete which enhances the properties of concrete. It was also stated that ultrafine particles less than 200 nm helps to reduce cement content and helps in the reduction of micro pores by acting as a filler agent and increase the density of concrete. Hosam M. Saleh et al. [22] discussed the formation of stabilized radioactive waste immobilization and construction materials form hazardous cement kiln dust. Portland cement with slag, silica fume, kiln dust along with 0.1% of nano materials were mixed and solidified. Compressive strength and porosity of the cement composite were found. It was reported that the performance with 0.1% nano silica increased the mechanical integrity by four fold. Hosam M. Saleh et al. [23] explored the possibility of improving the properties of cement by adding iron slag and titanate nano fibers to stabilize the radio active waste. The mechanical and physical characterization of the cement was enhanced. It was reported that it captured radionuclides from the contaminated aqueous solution before the immobilization process. Hosam M. Saleh et al. [24] studied the performance of cement-slag-titanate nano fibers composite immobilized radioactive waste solution. It was observed that cement nano composite was created by mixing iron slag with Portland cement which was hydrated with aqueous titanate nano fibers. SEM, FT-IR and X-ray diffraction analysis was performed to confirm the calcium silicate hydrate formation. It was observed that nano composite enhanced the mechanical and durability properties of cement and cement based materials. The chemical stability of cement-slag-titanate nano fibers composite was studied by monitoring leaching of ^{137}Cs which confirms immobilization radioactive waste and other hazardous waste. Hosam M. Saleh et al. [25] studied the influence of severe climatic variability on the structural, mechanical and chemical stability of cement kiln dust slag nano silica composite. The dust from cement kiln dust was mixed with 20% of iron slag and 0.1% nano silica, to produce modified cementitious composites which are suitable for construction and waste stabilization applications. The leachability studies showed that the impact of flooding can be reduced. Further freezing and thawing studies showed that the immobilization of radioactive waste can be enhanced.

Maile Aiu and Huang [9] synthesized the components of portland cement type I nano particles using sol gel process and compared the properties with that of commercial cement. Scanning Electron Microscope study revealed that nano cement particles were of size between 40 nm to 100 nm and appeared to be conglomerated and spherical. Energy Dispersive X-Ray Analysis test showed that the calcium to silica ratio was 3:1 or 2:1 and X-ray powder diffraction (XRD) result showed that nano cement contains C_3S, C_2S and copper oxide. Thomas Paul et al. [12] explained about the preparation, characterization of nano structured fly ash. The class F fly ash was ground in a high energy ball milling and converted into nano structured material. The nano structured fly ash was characterized for its particle size using particle size analyzer. Specific surface area was found using Brunauer-Emmet-Teller (BET) surface area apparatus. Fourier Transform Infrared Raman analysis (FTIR), SEM and TEM were used to study the particle aggregation and shape of the particles. On ball milling, the particle size got reduced from 60 μm to 148 nm i.e., by 405 times and the surface area increased from 0.249 m^2/g to 25.53 m^2/g i.e. by more than 100%. Surface of the nano structured fly ash was found to be more active as was evident from the FTIR studies. Morphological studies revealed that the surface of the nano structured fly ash was more uneven and rough and shape is irregular as compared to fresh fly ashes which are mostly spherical in shape. Narasimha Murthy Inampudi et al. [26] studied the crystallite size and lattice strain using XRD when micro sized fly ash was converted to nano sized fly ash using high energy ball milling process. The characterization was done after every 5 hours and at the end of 30 hours, the size of nano fly ash was found to be 83 nm. It was observed that the crystallite size was decreased and lattice strain was increased. It was also observed that the spherical shaped fly ash was converted to irregular shaped nano fly ash particle. The nano fly ash particles increased the hardness of composite and improved the compressive strength of cement composite. Gujjala Raghavendra et al. [27] presented the method of converting uneven micro size fly ash in to smooth glassy nano sized fly ash by planetary ball milling. From BET and XRD studies, it was noted that after 16 hours of milling, the surface area increased from 0.31m^2/g to 24.65m^2/g, the crystalline structure reduced from 59–26% and the particle size converted from 11 μm to 148 nm. Sada Abdal khaliq Hasan Alyasri et al. [28] investigated the economic feasibility of producing nano cement in a large scale through the cement factories. The mineral admixtures such as fly ash and slag should be added with crushed clinker where the moisture content should be maintained below 3%. The mixture was ground for 30 to 40 min to get nano cement. The specific gravity of nano cement was found to be 2.11 and specific surface area was 3,582,400cm^2/g. Bickbau and Shykun [29] explained that the Russian federal agency on technical regulating and metrology formulated the national standard on nano modified portland cement. Nano cement was produced by grinding process in ball mills of the clinker. Silicate minerals and gypsum were also added. The properties of nano cement were checked for its consistency. The specific surface area found by Blaine's apparatus should be below 400m^2/kg and grain size should be 10-100 nm. The mineral supplements were added in clinker to get economical nano cement, to reduce cost of fuel, reduce CO_2 emissions and to improve the quality of concrete with nano cement.

Parang Sabdono et al. [30] studied the effect of nano cement on the compressive strength of cement mortar. It was proved that use of nano cement in mortar helps in the reduction of the micro voids present in cement mortar, enhances the pozzolanic activity and increase the hydration rate of cement. Ordinary Portland Cement (OPC) and Portland Pozzolana Cement (PPC) were converted to nano particles of size 50 nm and 47 nm size. The experimental result showed that the initial setting time reduced from 138 min to 75 min with nano OPC and 123 min to 45 min with nano PPC. The compression test was conducted on 28 days cement mortar specimens of size 50 mm x 50 mm x50mm. The compressive strengths of mortar with nano OPC and nano PPC were found to be 68.493 N/mm^2 and 65.286 N/mm^2 respectively. It was reported that the nano cement improved the hydration reaction, lowered the initial setting time and increased the compressive strength. Ikhlef Bualem [31] studied the properties of cement and cement mortar with two combination of nano particles. First combination used was grinding 100% OPC to nano size without mineral additives and the second combination was grinding 50% of OPC together with 50% of silica sand (it is equal parts of granulated blast furnace slag and quartz sand) to nano size. It was found that the specific surface area of nano cement without mineral additive was 4900 cm^2/g and nano cement with 50% of mineral additive, the specific surface area was found to be 5000 cm^2/g. It was also found that the 90th day compressive strength of nano cement mortar was 51.4 MPa. for cement mortar with 100% nano cement, the strength was 114.3 MPa and for mortar with nano cement and mineral additive the strength was 77.8 MPa.

Gengying Li [32] compared the properties of normal cement concrete, high-volume fly ash high-strength concrete (SHFAC) incorporating nano silica and high-volume fly ash high-strength concrete (HFAC). Compressive strength was found from 3 days to 2 years. It was found that HFAC showed a 10% less compressive strength upto 56 days compared to normal cement concrete but increased to 21% higher than that of normal cement concrete at 2 years. The compressive strength of SHFAC showed that addition of 4% nano silica helped to gain early age strength by about 81% and also helped in gaining later strength upto 47% compared with normal cement concrete. The pore size of SHFAC was found to be less than that of NCC and HFAC. Zaki and KhaledRagab [33] studied the influence of nano silica on the properties of fresh and hardened normal cement concrete. In this study 18% of silica fume and 0.5%, 0.7% and 1% of nano silica were used. It was found that the workability of concrete improved with addition of super plasticizers when nano silica was added. It was reported that concrete with nano silica had a higher compressive strength, since nano silica not only acts as filler but helps in rapid formation of C-S-H gel. It was also reported that the efficiency of nano silica depends on its morphology, size and method of production. It was stated that 0.5% nano silica was found to perform better with and without silica fume compared to normal cement concrete. Nili et al. [34] discussed the performance of concrete with using nano silica and micro silica. It was reported that due to the high pozzolanic property, the compressive strength of concrete with 1.5, 3, and 4.5% of nano silica and 3, 4.5, 6 and 7.5% of micro silica gave a higher value than the normal cement concrete. It was found that the compressive strength of concrete with 1.5% nano silica and 6% of micro silica gave optimum values. Praveen and Janagan [35] experimentally studied the strength of concrete with nano particles. In this study, cement was replaced with 30% nano fly ash along with 3% nano Ground Granulated Blast Furnace slag (GGBS), 40% nano fly ash along with 4% nano GGBS and 50% nano fly ash along with 5% nano GGBS. The compressive strength and tensile strength of concrete were investigated on 7th and 28th day. It was reported that the strength was more when cement was replaced with 30% of nano fly ash along with 3% nano GGBS. Harihanandh and Sivaraja [36] experimentally studied the compressive strength, tensile strength and flexural strength of M20 grade concrete with nano fly ash. In this study class F calcium fly ash was converted into nano size in a ball grinding mill and its size was confirmed by SEM analysis. It was found that concrete with cement replaced with 23% nano fly ash gave 34% more compressive strength, 58.57% more tensile strength and 33.07% more flexural strength than the normal cement concrete. It was reported that the nano fly ash filled the pores and made the concrete denser.

It is obvious from the review of literature, that the application of nanotechnology in concrete is one of the promising areas of research. Studies were done on cement paste, cement mortar and cement concrete with incorporation of small percentage of nano materials. Nano particles of silicon dioxide, cement, fly ash, clay, metakaolin, iron oxide, aluminum oxide, zirconium oxide, carbon nano tubes, carbon nano fibers and nano carbonates are considered by many researchers in which nano silica was commonly used. Syntheses of nano materials were done by various methods and the properties of the nano materials were found to be dependent on the method of production. Characterization is done to understand the property of nano materials. Since the particle is of nano size with high specific surface area, the behaviour of cement paste, mortar and concrete with nano materials may be different from that of the normal cement paste, mortar and concrete. Normal consistency and setting time tests are usually carried out with cement paste, strength and durability studies are done with mortar and concrete. Few studies were also carried out with the addition of admixtures.

The special focus of the present investigation intends to highlight the replacement of cement with nano cement in larger percentages. Though many methods of production of nano materials are given in literature, ball milling method is adopted by which a large amount of nano materials can be produced economically without chemical change. Hence an attempt has been made to study the properties of nano cement, properties of cement paste with nano cement, properties of cement mortar with nano cement and compressive strength of concrete with nano cement.

3. Properties of materials

Concrete is considered to be a composite material containing a binder medium within which aggregate particles are embedded. The properties of materials used in the experimental investigation are cement, fine aggregate, coarse aggregate, and water. The properties are presented in this session.

3.1 Cement

Ordinary Portland Cement (OPC) of 53 grade conforming to IS: 12269-2013 [37] and procured from a single source was used for this investigation. The chemical and physical properties of the cement used are given in **Table 1**.

Particulars	Results (%)	Requirements of IS:12269
SiO_2	21.8	—
Al_2O_3	4.8	—
Fe_2O_3	3.8	—
CaO	63.3	—
SO_3	2.2	—
MgO	0.9	Maximum6
Na_2O	0.21	—
K_2O	0.46	—
Cl	0.04	Maximum0.1
P_2O_5	<0.04	—
Loss of ignition	2.0	Maximum4
Insoluble residue	0.4	Maximum3
Specific surface area, m^2/kg	370	Minimum225
Initial setting time, minutes	50	Minimum30
Final setting time, minutes	510	Maximum600
Standard consistency, %	34	—
Soundness, Le-chatelier, mm	1.0	Maximum10
Compressive strength, MPa		
3–days	42.5	Minimum27
7–days	48.0	Minimum37
28–days	63.5	Minimum53
Specific gravity	3.15	—

Table 1.
Chemical and physical properties of 53 grade OPC.

3.2 Fine aggregate

The locally available clean and dry natural sand from Cauvery river basin, Karur, India free from debris was used as fine aggregate. The specific gravity of fine aggregate was found to be 2.65. From sieve analysis, it was confirmed that the sand belongs to Zone II grading. Bulk density of fine aggregate was found to be 1520 kg/m^3. Fineness modulus of sand was found to be 2.32. The properties of fine aggregate were found to confirm with IS: 383-2016 [38].

3.3 Coarse aggregate

The coarse aggregate used was natural hard broken granite stones. Crushed granite metals of size 20 mm were used. The specific gravity of coarse aggregate was 2.79 and it was confirming to IS: 383-2016 [38].

3.4 Water

Potable water available in laboratory was used for casting and curing all specimens in this investigation. The quality of water was found to satisfy the requirements of IS: 456-2000 [39].

4. Synthesis and characterization of nano cement

4.1 Production of nano cement

Nano cement was produced by grinding ordinary Portland cement of 53grade in a high intensity ball grinder for 12 hours. In high energy ball grinding milling machine, high impact collisions were used to reduce microcrystalline materials down to nano crystalline structure without chemical change. Care was taken to avoid balling effect and agglomeration.

4.2 Microstructure analysis of nano cement

The particles size of the nano cement were found by surface morphology studies, the specific surface area was found by Brunauer Emmett Teller theory, and the elemental compositions were found by X-ray spectroscopic method.

4.2.1 Particle size determination of nano particles

Surface morphology studies were carried out using a Scanning Electron Microscope (JEOL, JSM 35 CF, Japan) shown in **Figure 1**.

53grade OPC was ground in the ball grinder mill for 12 hours to produce nano cement which was taken for morphological study using SEM. The SEM images of cement and nano cement are shown in **Figures 2** and **3** respectively.

From the SEM image, it can be seen that the cement particles have been ground to nano size. The size varies between 45 nm to 86 nm. It was also found that the shape of the particles was not altered due to grinding and agglomeration of cement particles did not take place.

4.2.2 Specific surface area of nano materials

The specific surface area is the total surface area of the exposed surface in square centimeter per unit mass. It was found by Brunauer, Emmett and Teller

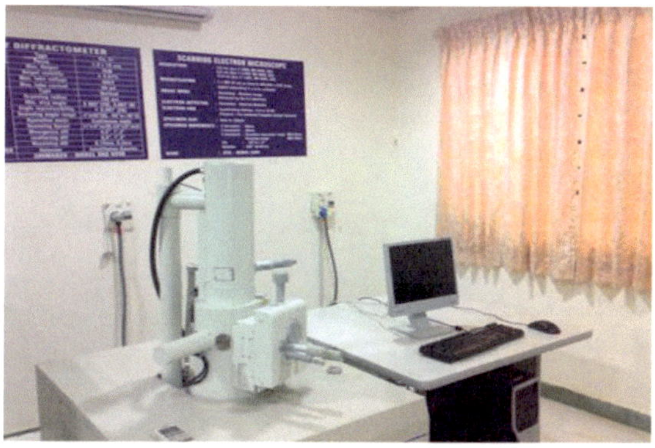

Figure 1.
Scanning electron microscope.

Figure 2.
SEM image of cement.

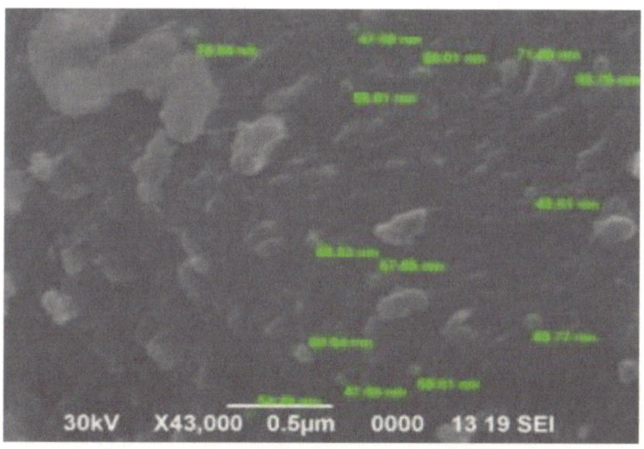

Figure 3.
SEM image of nano cement.

method for cement and nano cement. The specific surface area of cement was found to be 3700 cm^2/g and for nano cement, the specific surface area was found to be 485,000 cm^2/g. It can be seen that the particles in nano size have much higher specific surface area compared to particles in micro size.

4.2.3 Energy dispersive X-ray analysis of materials

EDAX is an x-ray spectroscopic method for determining elemental compositions. EDAX studies were carried out using a Scanning Electron Microscope. EDAX analysis

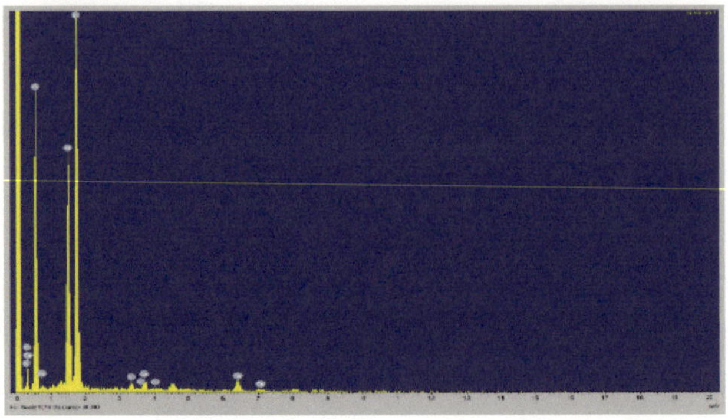

Figure 4.
EDAX of cement.

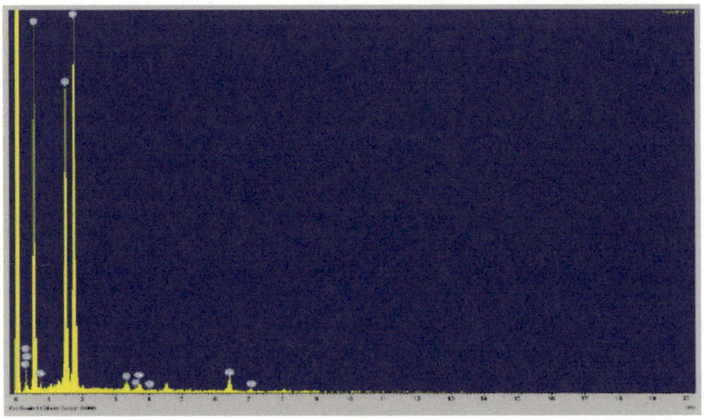

Figure 5.
EDAX of nano cement.

Element	Atomic weight percentage of chemical elements (%)						
	C	O	Al	Si	K	Ca	Fe
Cement	21.55	58.27	6.84	12.17	0.20	0.28	0.69
Nano Cement	19.80	59.95	8.00	11.07	0.24	0.19	0.75

Table 2.
EDAX analysis of particles.

was done in conjunction with SEM analysis. In EDAX analysis, X rays are emitted from the sample due to bombardment of electron beam from a spot, an area, a line profile or a 2D map. These X-rays are detected to characterize the elemental composition. In the EDAX images, the X axis represents energy and Y axis represents intensity. EDAX of cement and nano cement are shown in **Figures 4** and **5** respectively.

The details of EDAX analysis are given in **Table 2**.

From the EDAX study it was noted that the chemical composition of elements does not vary much when ground to nano particles.

5. Tests on cement paste with nano cement: normal consistency and setting time

The consistency test and setting time test on cement mortar with nano cement were carried out using the Vicat's apparatus conforming to IS: 5513-1996 [40].

Particulars	% Replacement of Cement with Nano Cement					
	0	10	20	30	40	50
Normal Consistency	34	33	33	34	35	33
Initial setting time	50	45	40	38	35	30

Table 3.
Normal consistency and setting time of cement paste with nano cement.

The values of consistency, initial setting time and final setting time of the cement paste with 0%, 10%, 20%, 30%, 40% and 50% of nano cement are given in **Table 3**.

From **Table 3**, it can be seen that the consistency of cement pastes with nano cement was almost the same but the initial and final setting times of are found to decrease as the replacement percentage increases.

The initial setting time of cement paste with nano cement was found to decrease to 30 minutes with 50% replacement whereas the initial setting time of normal cement paste is found to be 50 minutes. As per the IS 12269-2013 [37], the initial setting time of cement should not be less than 30mins. Hence the replacement level of cement with nano cement should not exceed 50%. The final setting time of cement paste with nano cement was found to decrease to 245 minutes at a replacement level of 50%.

6. Tests on cement mortar with nano materials

The compressive strength of cement mortar was determined on the 3rd, 7th, 21st and 28th days. After the 28th day test, the powder of the tested cement mortar cubes was taken for micro structural studies. The properties were evaluated by SEM, EDAX and FTIR test.

6.1 Compressive strength of cement mortar by experiment

The compressive strength of hardened mortar cubes of size 70.6 mm x 70.6 mm x 70.6 mm with and without nano cement were found using a compression testing machine of capacity 2000kN. The load was applied with a uniform rate of 35 N/mm^2/min after the specimen had been centered in the testing machine. The compressive strengths of cement mortar cubes are shown in **Table 4**.

The compressive strength of cement mortar was found to increase upto 50% replacement of cement with nano cement. The percentage increase in strength was found to vary between 34.36 and 77.85 for nano cement.

% replacement of cement with nano cement	Average Compressive Strength of Cement mortar (N/mm^2)			
	3^{rd}day	7^{th}day	21st day	28th day
0	42.50	48.00	57.00	63.50
10	58.98	74.48	83.23	85.32
20	62.98	79.50	88.42	90.66
30	65.50	81.20	90.96	93.24
40	67.88	85.37	92.44	95.68
50	70.00	75.53	94.00	98.00

Table 4.
Compressive strength of cement mortar with nano cement.

6.2 Microstructure of cement mortar with nano cement

Scanning Electron Microscope image of the crushed cement mortar particles cured for 28th days is shown in **Figure 6**.

It can be seen from **Figure 6** that the textures of particles consists of standalone clusters which indicates less formation of C-S-H gel. Hence the strength will be less than the mortar containing nano cement.

SEM images of cement mortar cubes in which cement was replaced with 10%, 20%, 30%, 40% and 50% of nano cement are shown in **Figure 7**.

From **Figure 7**, it can be seen that with 10% nano cement, the texture of hydrate particles are standalone clusters and with 50% nano cement, the texture of particles are colloidal due to the formation of C-S-H gel and hence the strength of mortar cubes with 50% nano cement is higher.

6.3 Energy dispersive X-ray spectroscopy study on cement mortar with nano materials

The EDAX study was carried out using the same sample used for SEM study. The EDAX of crushed cement mortar is shown in **Figure 8**.

The EDAX images of cement mortar in which cement was replaced with NC are shown in **Figure 9**.

From the EDAX analysis, the chemical elements present in the mortar with nano cement were found and the details are given in **Table 5**.

From the chemical composition, Ca/Si ratio was found. The strength of the mortar depends on Ca/Si ratio. Wolfgang Kunther et al. [41] reported that the strength of mortar cubes decreases as the Ca/Si ratio increases. It can be seen from **Table 5** that the ratio of Ca/Si decreases for replacement levels and hence maximum strength is obtained at 50% replacement level.

6.4 FTIR Spectrum for 28 days cement mortar with nano materials

Fourier transform infrared spectroscopy is a technique used to obtain an infrared spectrum of absorption or emission of a solid, liquid or gas. an Fourier transform infrared spectroscopy spectrometer simultaneously collects

Figure 6.
SEM image of cement mortar cured for 28th days.

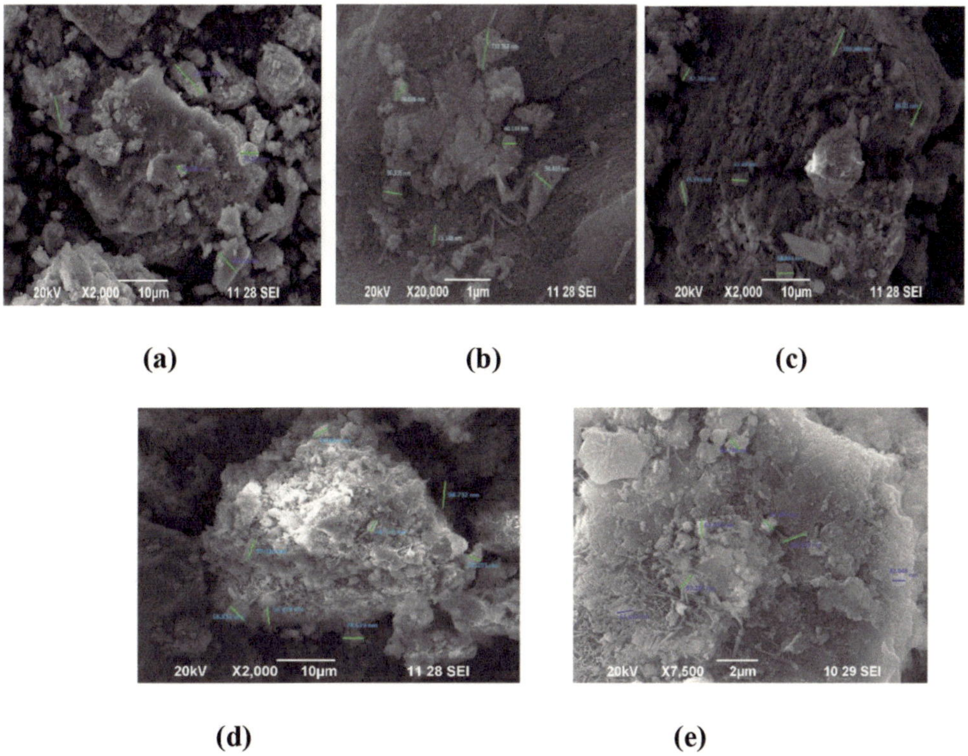

Figure 7.
SEM images of cement mortar cubes with nano cement cured for 28th days (a) 10% replacement (b) 20% replacement (c) 30% replacement (d) 40% replacement (e) 50% replacement.

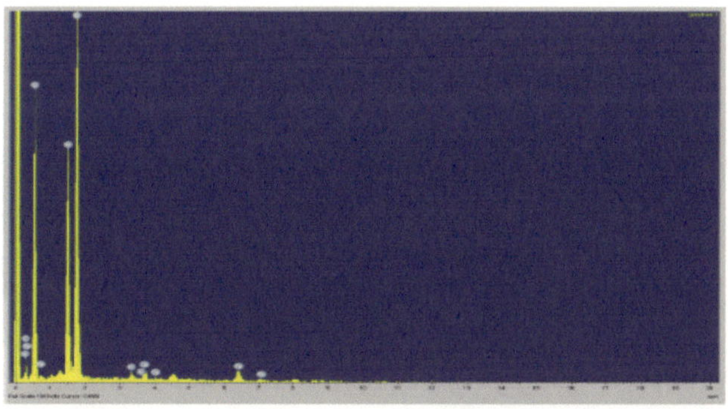

Figure 8.
EDAX image of cement mortar cube cured for 28th days.

high-spectral-resolution data over a wide spectral range. The an Fourier transform infrared spectroscopy spectrum was recorded on IR PRESTIGE 21, SHIMADZU spectrophotometer at ambient temperature using a KBr disk method. The resulting spectrum creates a molecular fingerprint of the sample representing the molecular absorption and transmission. The changes in the molecular groups in the cement mortar before and after addition of nano particles were made by a Fourier transform infrared spectroscopy analysis. In an Fourier transform infrared spectroscopy test, the X-axis of an IR spectrum is labeled as Wave number (1/cm) and ranges from 400 cm^{-1} on the far right to 4000 cm^{-1} on the far left.

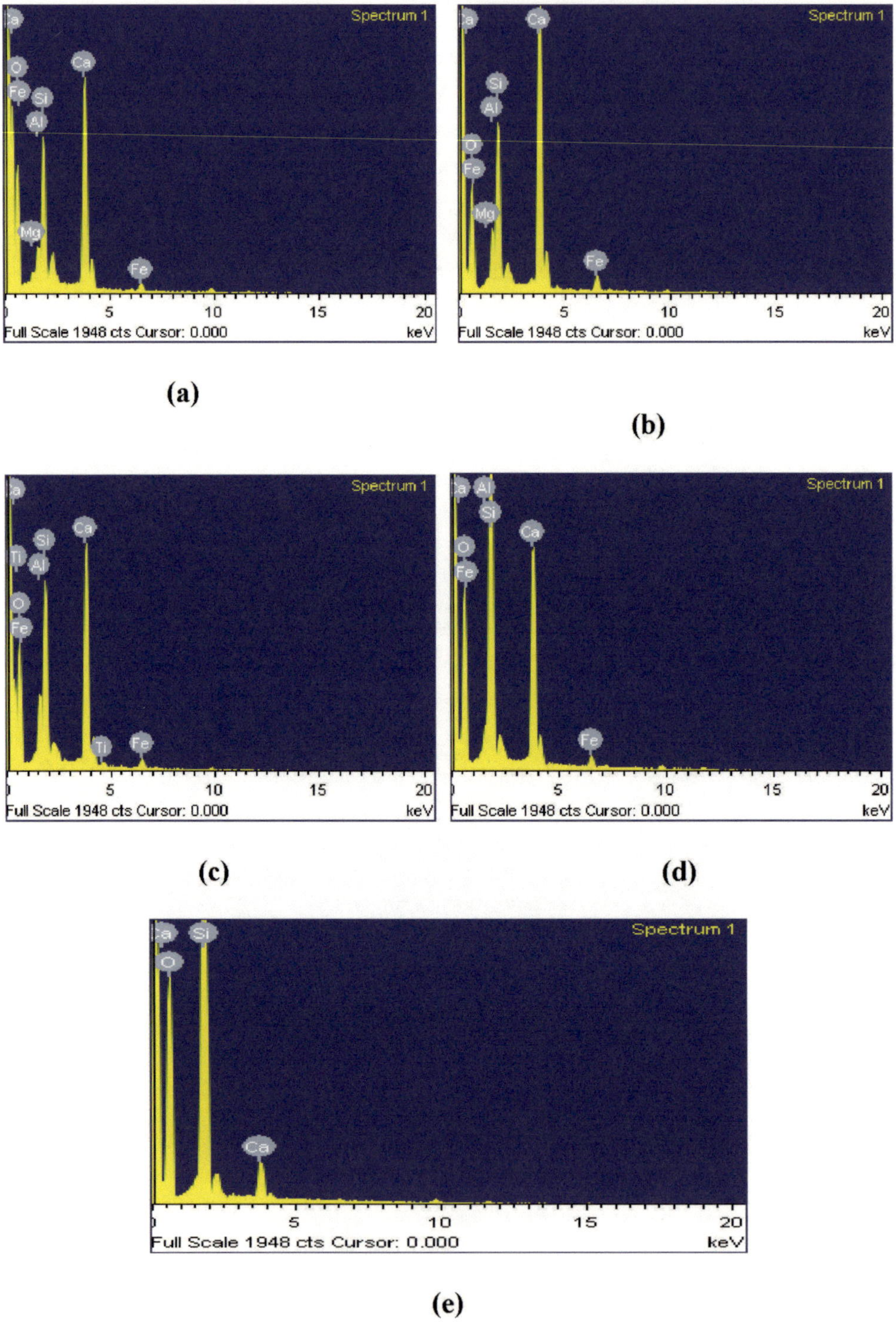

Figure 9.
EDAX images of cement mortar cubes with nano cement cured for 28th days (a) 10% replacement (b) 20% replacement (c) 30% replacement (d) 40% replacement (e) 50% replacement.

The Y-axis is labeled as Transmittance in % and ranges from 0 at the bottom to 100 at the top. The characteristic peaks in the infrared spectrum were determined. All infra-red spectra contain many peaks. However, the large peaks on the spectrum will provide the data necessary to read the spectrum. The regions of the spectrum in which the characteristic peaks exist were determined. The characteristic peak is compared to IR spectrum and the compounds were identified. an Fourier transform infrared spectroscopy transmission spectrum of the normal cement mortar is shown in **Figure 10**.

According to Hasan Biricika and Nihal Sarierb [42], Xu et al. [43] and Varas et al. [44], the band of the spectra between 3640 cm^{-1} to 3400 cm^{-1} corresponds to the structural -OH group formed during the hydration of C_2S and C_3S and free -OH group from water molecules present in the mixture. The spectral band between 2500 cm^{-1} and 1500 cm^{-1} corresponds to the H-O-H absorbed water molecule group which indicates the decrease in the free water and between 1500 cm^{-1} to 400 cm^{-1} and it corresponds to Si-O and Si-O-Si silicate group which indicate the formation C-S-H.

The FTIR transmission spectra of cement mortar in which cement was replaced with nano cement are shown in **Figure 11**. From the FTIR transmission spectra, the peaks attained by the mortar with nano cement are given in **Table 6**.

From **Figure 10** and **Table 6**, it can be seen that the band value of 3429.43 cm^{-1}diminishes to 462.92 cm^{-1} which implies the formation of C-S-H gel.

Figure 11 and **Table 6**, it can be seen that the band value between 3635.74 cm^{-1} and 3410.28 cm^{-1}dimnishes to 488.34 cm^{-1} and 455.87 cm^{-1} which implies that strength has been achieved. It can be seen that the lowest band value is seen in 50% replacement of cement with nano cement and the decrease in the band value implies a decrease in free water and increase in strength.

S.No	Element	Atomic percentage of elements					
		% Replacement of cement with nano cement					
		0%	10%	20%	30%	40%	50%
1	O	53.9	78.3	71.7	73.7	71.7	69.1
2	Al	5.80	2	3.3	3.9	1.5	0.43
3	Si	12.7	7.9	9.1	10.2	17.9	29.4
4	Ca	26.9	10.3	14.1	11.3	8.3	1.5
5	Fe	0.70	0.6	1	0.7	0.6	0.12
6	Ca/Si	2.12	1.3	1.5	1.1	0.5	0.1

Table 5.
Results of EDAX analysis of cement mortar with nano cement.

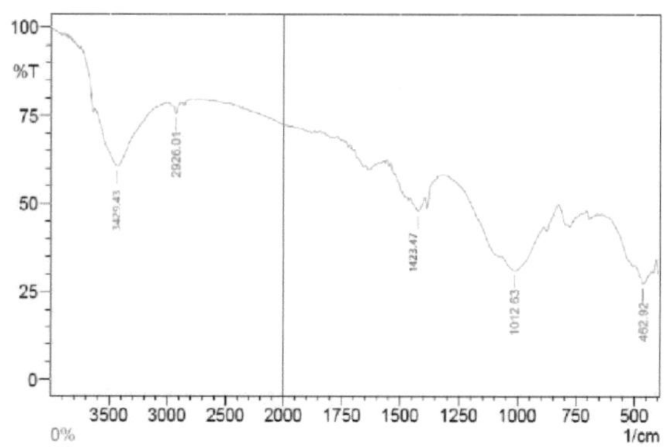

Figure 10.
FTIR transmission spectrum of cement mortar cured for 28th days.

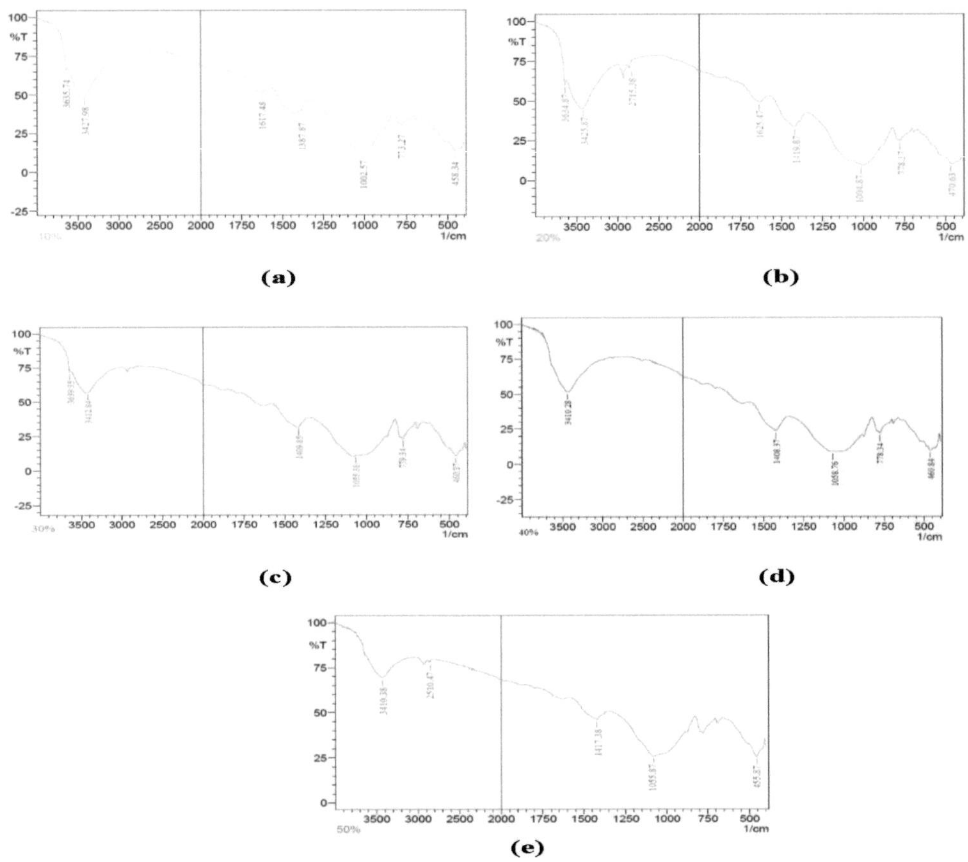

Figure 11.
FTIR transmission spectra of cement mortar on 28th day with nano cement (a) 10% replacement (b) 20% replacement (c) 30% replacement (d) 40% replacement (e) 50% replacement.

FTIR spectra peak of mortar with NC in cm^{-1}					
%RCNP					
0	**10**	**20**	**30**	**40**	**50**
3429.43	3635.74	3634.87	3639.35	3410.28	3410.28
2626.61	3427.98	3425.87	3412.84	1408.37	2510.47
1423.47	1617.48	2715.38	1409.85	1058.76	1417.38
1012.63	1387.87	1625.47	1055.38	778.24	1055.87
462.92	1002.57	1419.87	779.34	460.84	455.87
	773.27	1004.87	460.87		
	488.34	778.37			
		470.63			

Table 6.
FTIR transmission spectrum peak of cement mortar with nano cement.

From both the experimental and micro structural studies, it was noted that the compressive strength of cement mortar increases as the replacement level of cement with nano cement increases. From the SEM analysis, it was seen that the nano particles filled the pores and made the concrete denser. Additional formation of C-S-H gel can be seen from SEM analysis, reduction of Ca/Si ratio from EDAX analysis and the reduction of band spectra values from FTIR analysis which indicate the increase in strength of cement mortar with nano particles.

7. Compressive strength of concrete with nano materials

The compressive strength of M20, M30, M40 and M50 grades of concrete with and without nano cement are presented in **Table 7**.

Figure 12 shows the compressive strength of concrete with respect to the percentage replacement of cement with NC.

From the **Table 7** and **Figure 12**, it can be seen that the compressive strength increases with the increase in percentage replacement of cement with nano cement. It can be seen that the compressive strength increases up to 50% replacement of cement with nano cement. The correlation coefficient between the compressive strength and percentage replacement of cement with nano cement were found to be 0.9971, 0.9946, 0.9717 and 0.9928 for M20, M30, M40 and M50 grades of concrete respectively.

It can be seen that the compressive strength increases as the replacement of cement with nano materials, curing days and grade of concrete increase. The increase in compressive strength was found to range between 0.71% and 31.5%. The percentage increase in the compressive strength of M20 grade was found to be more than that of M50 grade of concrete. Saloma et al. [45] reported that the rapid development of the compressive strength of concrete with nano materials is due to the fact that nano materials serve as a filler to increase the density and as an activator

Average compressive strength of concrete with nano cement N/mm^2						
GC		**M20**			**M30**	
%RCNC	**28th day**	**56th day**	**90th day**	**28th day**	**56th day**	**90th day**
0	25	27.3	33.1	32	34.7	42
10	26.1	28.7	36.5	32.6	35.6	43.4
20	27.6	29.9	37.3	33.7	36.7	44.6
30	28.4	32.2	38.9	35.1	37.9	46.3
40	29.5	34	41.1	36.4	38.9	47.7
50	31	35.9	42.8	37.4	39.7	48.8
GC		**M40**			**M50**	
%RCNC	**28th day**	**56th day**	**90th day**	**28th day**	**56th day**	**90th day**
0	42	44.3	48	49.9	51.4	59.3
10	44.3	46.6	49.6	52.3	55.1	62.6
20	46	47.6	52.7	53.2	56.7	64.2
30	45.7	48.3	54	55.3	58.6	66.8
40	47.8	49.8	55.5	57.2	60.9	69.4
50	49.2	50.8	56.6	59.9	63.9	70.9

GC is grade of concrete.
%RCNC is percentage replacement of cement with nano cement.

Table 7.
Compressive strength of concrete with nano cement.

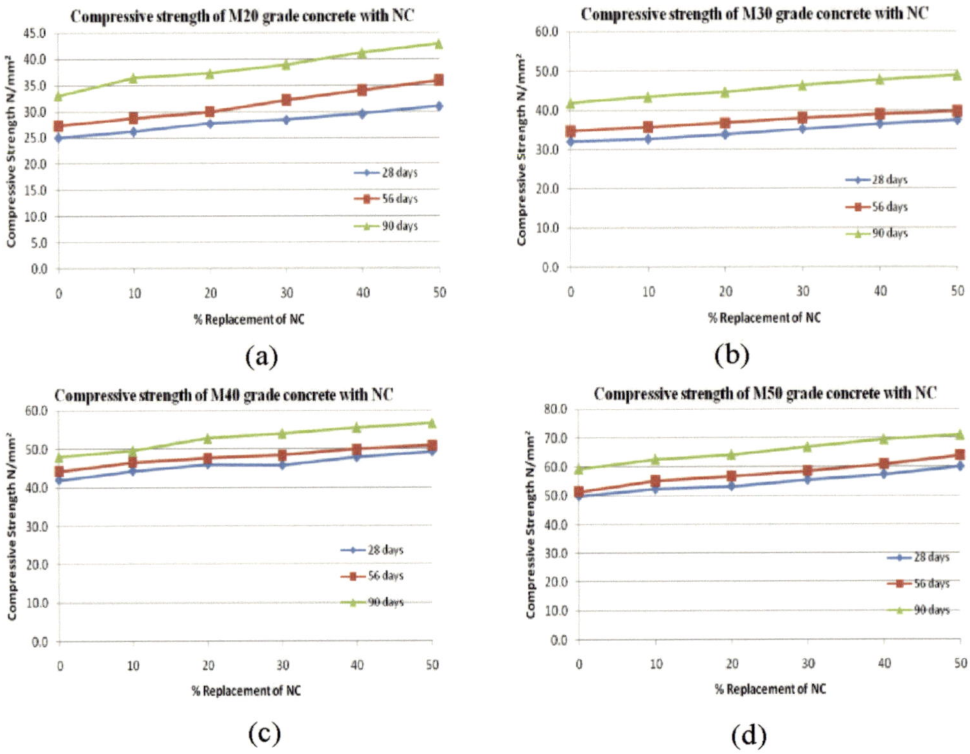

Figure 12.
Compressive strength of concrete with nano cement(a)M 20 grade of concrete (b)M 30 grade of concrete (c)M 40 grade of concrete (d) M 50 grade of concrete.

NP	Particulars	Variations in compressive strength			
		M 20	M30	M40	M50
NC	%	4.4 to 31.5	1.88 to 16.88	3.33 to 17.92	4.81 to 24.32
	ratio	1.04 to 1.32	1.02 to 1.17	1.03 to 1.18	1.06 to 1.24

Table 8.
Range in the variations in compressive strength for different grades.

in the hydration reaction and reacts with free $Ca(OH)_2$resulting in concrete with higher compressive strength. The ranges in the variations of compressive strength are given in **Table 8**.

From the **Table 8**, it can be seen that the nano cement is very effective in increasing the compressive strength of concrete.

8. Conclusions

8.1 Effect of grinding on materials

The micro-sized cement was converted to nano size by grinding it in a ball grinding mill for 12 hours and the particle size was found to range from 45 nm to 86 nm.

The specific surface areas of nano cement, increased by 13008.11%, when compared with that of ordinary portland cement.

The chemical properties of nano sized particles were found to be the same as the particles before grinding.

8.2 Effect of nano materials on properties of cement paste and cement mortar

The normal consistency of cement paste with nano cement was found to range between 33% and 35%.

The initial setting time of cement paste with nano cement, was found to decrease to 30 minutes at 50% replacement level when compared to the initial setting time was 50 minutes for the normal cement paste.

The final setting time of cement paste with NC, was found to decrease to 245 minutes at 50% replacement level when compared with the final setting time of510 minutes for the normal cement paste.

The percentage increase in compressive strength of cement mortar with nano cement was found to range between 23.76 and 64.91 when compared with the compressive strength of the normal cement mortar.

The optimum replacement level of nano cement is 50%. Replacement level beyond 50% will result in the rapid setting which is not desirable.

8.3 Effect on compressive strength of concrete

The compressive strength of concrete was found to increase as the replacement level of cement with nano cement increases for all grades of concrete and for all curing days considered. The percentage increase in strength was found to vary between 2% and 29.3% for nano cement. The lower percentage of 2% was obtained for M30 mix cured for 28 days at a replacement percentage of 10%. The higher percentage of 29.3% was obtained for M20 mix cured for 90 days at a replacement percentage of 50%.

Conflict of interest

The authors declare no conflict of interest.

Author details

Jemimah Carmichael Milton[1*] and Prince Arulraj Gnanaraj[2]

1 Department of Civil Engineering, Vignan's Lara Institute of Technology and Sciences, Vadlamudi, Guntur, Andra Pradesh, India

2 School of Engineering and Technology, Karunya Institute of Technology and Sciences, Coimbatore, Tamil Nadu, India

*Address all correspondence to: jemimahcarmichael@gmail.com

References

[1] Konstantin Sobolev. Modern developments related to nanotechnology and nanoengineering of concrete. Frontiers of Structural and Civil Engineering. 2016:**10**(2):131-141

[2] Hanus MJ. Nanotechnology innovations for the construction industry. Progress in materials science. 2008

[3] Gann D. A Review of Nanotechnology and its Potential Applications for Construction. SPRU, University of Sussex. 2002

[4] Perumalsamy B, Ken C. Nanotechnology of concrete: Recent developments and future perspectives, Nanotechnology and Concrete: Proceedings of ACI Session; 2006:15-28

[5] Surinder M. Nanotechnology and construction. European Nanotechnology Gateway-Nanoforum Report, Institute of Nanotechnology; 2006:2-10

[6] Florence S, Konstantin S. Nanotechnology in concrete–a review. Construction and building materials; 2010;24:2060-2071

[7] Bhuvaneshwari. B, Saptarshi S, Nagesh RI. Nanoscience to Nanotechnology for Civil Engineering - Proof of Concepts. Proceedings of the 4th WSEAS International conference on Energy and development-environment-biomedicine, Corfu Island, Greece. 2011:230-235

[8] Hui L, Hui-gang X, Jie Y, Jinping O. Microstructure of cement mortar with nano-particles. Composites Part B Engineering; 2004;**35**:185-189

[9] Maile A, Huang CP. The chemistry and physics of nano-cement. Report submitted to NSF REU University of Delaware; 2006:1-28

[10] Tao J. Preliminary study on the water permeability and microstructure of concrete incorporating nano-SiO2. Cement and concrete research; 2005;**35**(10):1943-1947

[11] Ali N, Shadi R. The effects of SiO2 nano particles on physical and mechanical properties of high strength compacting concrete. Composites Part B: Engineering; 2011;**42**(3):570-578

[12] Thomas PK, Satpathy SK, Manna I, Chakraborty KK, Nando GB. Preparation and characterization of nano structured materials from fly ash. Nanoscale Research Letters; 2007;**2**(8):397-404

[13] Hanus MJ, Andrew TH. Nanotechnology innovations for the construction industry. Progress in materials science; 2013;**58**(7):1056-1102

[14] Silverstre J, Silvestre N, Brito JD. Review on concrete nanotechnology. European Journal of Environmental and Civil Engineering; 2016;**20**(4):455-485

[15] Elzbieta H. Properties of cement-based composite modified with magnetic nanoparticles: A review Materials; 2019:1-34

[16] Konstantain S, Miguel FG. How nanotechnology can change the concrete world. American ceramic Society Bulletin; 2005;**84**:16-20

[17] Zhi G, Zhili G. Applications of nanotechnology and nanomaterials in construction. First International Conference on Construction in Developing Countries (ICCIDC–I), Advancing and Integrating Construction Education, Research & Practice, Karachi, Pakistan; 2008:235-240

[18] Laila R, James B, Rouhollah A, Jon M, Taijiro S. Cement and concrete nanoscience and nanotechnology. Materials; 2010;**3**:918-942

[19] Jafarbeglou M, Abdouss M, Ramezanianpour AA. Nanoscience and nano engineering in concrete advances a review. International journal of nano science and nanotechnology; 2015;**11**(4):263-273

[20] Guillermo B, Faustino P, Faustino P, Julia A. Nano-inclusions applied in cement-matrix composites: A review. Materials; 2016;**9**(12):10-15

[21] Muhd NMS, Hamidah MS, Mohd FA. Applications of using nano materials in concrete: A review, Construction and Building Materials; 2017;**133**:91-97

[22] Hosam MS, Fathy A El-Saied, Taher AS, Aya AH. Macro-and nanomaterials for improvement of mechanical and physical properties of cement kiln dust-based composite materials. Journal of Cleaner Production; 2018a-1;**204**:532-541

[23] Hosam M, El-Sheikh SM, Elshereafy EE, Essa AK. Mechanical and physical characterization of cement reinforced by iron slag and titanate nano ibers to produce advanced containment for radioactive waste. Construction and Building Materials; 2018b -2; **200**:135-145

[24] Hosam MS, El-Saied FA, Salaheldin TA, Hezo AA. Influence of severe climatic variability on the structural, mechanical and chemical stability of cement kiln dust-slag-nanosilica composite used for radwaste solidification. Construction and Building Materials; 2019a -1;**218**:556-567

[25] Hosam MS, El-Sheikh SM, Elshereafy EE, Essa AK. Performance of cement-slag-titanate nanofibers composite immobilized radioactive waste solution through frost and flooding events. Construction and Building Materials; 2019b -2; **223**:221-232

[26] Narasimha MI, Venkata RD, Babu Rao J. Microstructure and mechanical properties of aluminum-fly ash nano composites made by ultrasonic method. Materials and design; 2011;**25**:55-65

[27] Gujjala R, Shakuntala O, Samir KA, Pal Sk. Fabrication and characterization of nano fly ash by planetary ball milling. International journal of material science innovations; 2014;**2**(3):59-68

[28] Sada AKHA, Iyad SA, Prabir KS. Feasibility of producing nano cement in a traditional cement factory in Iraq. Case studies in construction materials. 2017;**7**:91-101

[29] Bickbau MY, Shykun VN. Nano-cements future of world cement industry and concrete technology. Program International conference, seminar, Dubai; 2017:3-33

[30] Parang S, Frisky S, Dion AF. The effect of nano-cement content to the compressive strength of mortar. 2nd International Conference on Sustainable Civil Engineering Structures and Construction Materials 2014 (SCESCM 2014), Procedia Engineering; 2014:386-395

[31] Ikhlef B. Test on nano cement mortar and concrete. Program International conference, Dubai; 2017:38-55

[32] Gengying L. Properties of high-volume fly ash concrete incorporating nano-SiO2. Cement and Concrete Composites; 2004;**34**(6):1043-1049

[33] Zaki SI, Khaled SR. How nano technology can change the concrete industry. SBEIDCOH International conference on sustainable Built Environment Infrastructures in

developing countries, Oran, Algeria; 2009:407-414

[34] Nili M, Ehsani A, Shabani K. Influence of nano-SiO2 and microsilica on concrete performance. Construction Materials and Technologies; 2010:1-7

[35] Praveen S, Janagan SS. Partial replacement of cement with nano flyash(class c) and nano GGBS. International research journal of engineering and technology; 2015;**2**(8):979-983

[36] Harihanandh M, Sivaraja M. Strength and mechanical properties of nano fly ash concrete. International Journal of Advanced Engineering Technology; 2016;**7**(2):596-598

[37] IS: 12269-2013. Indian Standard Specification for 53 Grade Ordinary Portland Cement. Bureau of Indian Standards, New Delhi, India

[38] IS: 383-2016. Indian Standard Specification for Coarse and Fine Aggregates from Natural Sources for Concrete. Bureau of Indian Standards, New Delhi, India

[39] IS: 456-2000. Indian Standard Code of Practice for Plain and Reinforced Concrete. Bureau of Indian Standards, New Delhi, India

[40] IS: 5513-1996. Vicat Apparatus – Specification. Bureau of Indian Standards, New Delhi, India

[41] Wolfgang K, Sergio F, Jorgen S. Influence of the Ca/Si ratio on the compressive strength of cementitious calcium–silicate– hydrate binders. Journal of Materials Chemistry; 2017;**33**:17401-17412

[42] Hasan B, Nihal S. Comparative study of the characteristics of nano silica, silica fume and fly ash incorporated cement mortars. Materials Research; 2014;**17**(3):570-582

[43] Xu P, Kikpatrick RJ, Poe B, McMillan PF, Cong X. Structure of calcium silicate hydrate (C-S-H): Near-, mid-, and far-infrared spectroscopy. Journal of the American Ceramic Society; 1999;**82**(3):742-748

[44] Varas MJ, Alvarez de Buergo M, Fort R. Natural cement as the precursor of Portland cement: Methodology for its identification. Cement and Concrete Research; 2005;**35**:2055-2065

[45] Saloma AN, Iswandi I, Mikrajuddin A. Improvement of concrete durability by nanomaterials. Procedia Engineering; 2015;**125**:608-612

Chapter 3

The Resistance of New Kind of High-Strength Cement after 5 Years Exposure to Sulfate Solution

Michal Bačuvčík, Pavel Martauz, Ivan Janotka and Branislav Cvopa

Abstract

This article deals with the determination of technically important properties, the recognition of microstructure and pore structure, and the mortar resistance of a new cement kind NONRIVAL CEM I 52.5 N containing 7.94% wt. of C_3A to 5% sodium sulfate solution. Both reference types of cement were industrially manufactured: 1) ordinary Portland cement CEM I 42.5 R and 2) Portland cement CEM I 42.5 R – SR 0, declared as sulfate resistant because of C3A = 0%. The research was carried out at standardized mortars. The used sodium sulfate solution, which contained 33802.8 mg of aggressive SO_4^{2-} per liter, exceeded approximately 5 to 10 times the concentration of the third degree of aggressiveness of the XA chemical environment according to STN EN 206 + A1. The reference medium was drinking water. The 5-year results of non-destructive and destructive physical-mechanical tests as well as the formed microstructure and pore structure in both liquid media were evaluated. The cause of the NONRIVAL CEM I 52.5 N sulfate resistance was explained, despite the manufacturer's declared C_3A content of up to 8% by weight. Sulfate resistance of NONRIVAL CEM I 52.5 N is found comparable to that of sulfate resistant CEM I 42.5 R – SR 0.

Keywords: high-strength cement, sulfate resistance, durability

1. Introduction

A sulfate attack represents one of the most aggressive ways of acting on concrete, which worsens the durability of the structures. There is a large number of civil engineering structures, such as the foundations of pillars, bridges or concrete canals, etc., which could be exposed to aggressive sulfates throughout their lifetime [1–3]. The resistance of concrete is increased by using durable types of cement compared to Portland, such as pozzolanic cement when natural or industrial pozzolans are added. Považská cementáreň a. s., Ladce has developed a new type Portland cement, designated as NONRIVAL CEM I 52.5 N, which does not meet the criteria for sulfate-resistant cement according to the requirements of STN EN 197–1 [4]. NONRIVAL CEM I 52.5 N is considered to be an innovative new generation cement [5] with a content of up to 5% wt. of industrially produced submicron-sized pozzolanic addition with minimum of 50% SiO2. Studies show that the response of a cement-based material to sulfate attack varies in many cases and is influenced by many factors. Most experiments took place on a macroscopic scale, but it should be noted that the essence of the resulting attack lies in changes in the microstructure and pore structure of the cement matrix. Therefore, it is necessary to study the sulfate attack on cement-based mortar not only by assessing its physical properties but also by analyzing a microstructure and pore structure [6]. The condition of the pore structure of concrete is an important criterion for assessing sulfate resistance as its strength, as it determines the permeability for the penetration of aggressive solution into the interior of the microstructure formed over time.

According to the source of sulfate ions, sulfate attack is divided into two main types by secondary ettringite formation: external and internal. External sulfate attack occurs when the source of sulfates comes from the external environment when sulfates penetrate the concrete structure. Internal sulfate attack is caused by internal sulfate sources in an environment without external sulfate sources, such as coming from aggregates or by the thermal decomposition of ettringite [7].

External sulfate attack, also referred to as traditional, is characterized by the chemical interaction of sulfate-rich soil or water with the hydrated cement matrix. Soils containing sodium, potassium, magnesium, and calcium sulfates are the main sources of sulfate ions in groundwater. Another source of sulfates is industrial wastewater, e. g. from the chemical or agricultural industry [8–10].

External sulfate attack occurs if the following three factors coexist:

a. high permeability of the cement composite/concrete structure;

b. sulfate-rich environment;

c. the presence of water.

The first step of the external sulfate attack is the penetration of sulfate ions from the outer environment into the concrete. Consequently, the transformation of calcium hydroxide and/or calcium silicate hydrate (C-S-H) to gypsum takes place according to specific reactions [11]. This process causes the hydrated cement matrix to expand, crack, and peel. Gypsum is prevailingly formed by the reaction of sulfate ions with $Ca(OH)_2$ or calcium silicate hydrate (C-S-H). However, a more important manifestation of sulfate attack is the reduction of the strength and cohesiveness of the developed cement matrix by the decalcination of C-S-H.

The STN EN 197–1 [4] defines sulfate-resistant cement for general use as a cement whose properties meet the requirements for sulfate resistance. Additional requirements for sulfate-resistant cement are the sulfate content (as SO_3) in cement, the C_3A content in clinker, and the pozzolanity of cement. The seven sulfate-resistant cements for general use are divided into 3 main types as follows:

1. Sulfate-resistant Portland cement with different C_3A content in clinker:

   - C_3A content in the clinker = 0% by weight, designated as CEM I-SR 0

   - C_3A content in the clinker $\leq$3% by weight, designated CEM I-SR 3

   - C_3A content in the clinker $\leq$5% by weight designated CEM I-SR 5;

2. Sulfate-resistant blast furnace cement without the requirement for C_3A content in clinker: either designated CEM III/B-SR or CEM III/C-SR;

3. Sulfate-resistant pozzolanic cement with C_3A content in a clinker ≤9% by weight: designated either CEM IV/A-SR or CEM IV/B-SR.

Sulfate-resistant types of cement, defined by the European standards [12, 13] must therefore meet the criteria for C_3A, SO_3, and pozzolan content, without additional requirements for verifying their resistance. The standards discriminate "new generation cements", which, despite not meeting the criteria in the defined standards, can show high sulfate resistance. Other ways to increase sulfate resistance are "innovative cement kinds" either hybrid cement [14] or as in this case, high-fine pozzolanic addition present in the cement. Such a cementitious composition with active submicron-sized pozzolanic particles forms a dense microstructure, poor in $Ca(OH)_2$ but rich in gel hydration products that are specified by low-permeable pore structure.

There are various ways to verify chemical resistance; a) long-term, multi-year exposures to aggressive media [15, 16] and b) accelerated [17, 18]. However, there is no known, unified and universally valid testing methodology worldwide. Therefore, the efficiencies of determinations are not mutually comparable; in addition, the same material systems and exposure conditions are not always tested. The final output of each method is a knowledge on the increased chemical (e. g. sulfate) resistance of the verified cement system compared to the reference. It is still problematic to determine the coherence between the laboratory test results (a or b) and the actual resistance time of concrete used in the field for decades of years either in aggressive soil or groundwater.

The objective of this chapter is to characterize the sulfate resistance of submicron-sized pozzolan containing NONRIVAL CEM I 52,5 N with up to 8% wt. C_3A and to explain the cause for its comparability with sulfate-resistant CEM I 42.5 SR 0 with none of C_3A.

2. Experimental procedure

2.1 Materials

Ordinary Portland cement (CEM I 42.5 N) as a reference cement 1 (PC), sulfate-resistant CEM I 42.5 R - SR 0 as a reference cement 2 (SR) and NONRIVAL CEM I 52.5 N as experimental cement (N), were used. Both reference types of cement were produced according to STN EN 197–1 [4] and NONRIVAL CEM I 52.5 N was prepared according to the internal cement plant's standard and SK technical assessment.

2.2 Casting and curing

Mortar specimens of size (40 × 40 × 160) mm with the cement to standard sand weight ratio of 1: 3 and water to cement ratio of 0.5, were prepared. The mortars were cured 24 hours at 20°C/95% R.H.-air in a climate chamber. After demolding, they were kept 27 days in water at (20 ± 1) °C (basic curing – BC), and then either in water (reference medium) and aggressive 5% sodium sulfate solution (33,800 mg SO_3 per liter) for 5-year exposure, respectively.

The tests of chemical resistance were conducted by the own methodology of "partially accelerated tests" [14] based on keeping the mortars in strongly over-concentrated aggressive solutions for a sufficiently long time. The aggressive environment was specified in the following way: every 1 cm^2 of the exposed area of prism must be in permanent contact with at least 10 cm^3 of 5% wt. Na_2SO_4. A sulfate solution and reference water were refreshed every 30 days within 90 days of testing, every 45 days between 90 and 365 days, and every 60 days up to 5 years of exposure, respectively.

2.3 Testing procedures for cement

All types of cement were tested for chemical composition by STN EN 196–2 [19]; consequently, the Bogue mineral composition was determined. Standard consistency, initial and final set, and soundness were verified by STN EN 196–3 [20]. After 2- and 28-day cure, flexural and compressive strengths of the mortars were obtained according to STN EN 196–1 [21].

2.4 Testing procedures for mortars conducted to sulfate attack

The consistency according to STN EN 1015–3 [22] represents the value of the degree of pouring of the formed fresh mortar after 15 strokes of the compaction table. The bulk density was determined in one-liter container according to STN EN 1015–6 [23]. Based on the consistency results, all mortars fall into the category of plastic mortar, according to STN EN 1015–6. Air content in the mortar was determined by the pressure method according to STN EN 1015–7 [24].

The hardened mortars were during 5 years of sulfate exposure continuously tested for length changes [25], dynamic modulus of elasticity (DME) [26], and periodically for flexural and compressive strength [21]. After destructive tests, the microstructure and pore structure were identified by X-ray diffraction analysis (XRD), thermal analysis (TG-DTA), mercury intrusion porosimetry (MIP), and scanning electron microscopy (SEM) techniques. The grounded mortars were sieved through a 0.063 mm mesh to receive the powder suitable for testing. For the XRD, the Philips diffractometer was used in a 2Θ range of 5–65°. CuKα radiation and Ni - filter was applied. Thermal analysis was performed on the Netzsch apparatus STA 449 F3 Jupiter in the air at a heating rate of 10°C/min. Basic parameters of the pore structure were identified by MIP using the high-pressure porosimeter Quantachrome Poremaster 60 GT. The JEOL 7500F device was used to study microstructure by scanning electron microscopy. Chemical composition, with special emphasis on the SO_3 content bound in the cement matrix, was estimated by the analytical procedures given in STN EN 196–2 [19].

3. Results and discussion

3.1 Basic properties of cements and mortars after basic curing

Chemical composition of the cements (N – NONRIVAL, SR – sulfate-resistant, and PC – reference Portland) is listed in **Tables 1** and **2**. The content of chloride ions 0.09% wt. in N-cement, 0.07% wt. in SR cement, and 0.06% wt. in PC-cement, is almost the same. The mineral composition was calculated by the Bogue formulas (**Table 2**).

All types of cement meet the requirements for chemical properties, which are given as characteristic values in STN EN 197–1 [4] based on a loss on ignition (LOI), which is less than 5% by weight, an insoluble residue, which is less than 5% by weight, a sulfate content (expressed as SO_3) of less than 4% by weight and a chloride content of less than 0.10% wt.

Cement	LOI (% wt.)	Ins. res. (% wt.)	Content of the component (% wt.)						
			SiO_2	CaO	Al_2O_3	Fe_2O_3	MgO	SO_3	Na_2O eq.
N	1.9	4.06	19.9	59.9	4.95	3.06	1.3	3.43	1.2
SR	3.5	2.22	21.3	61.0	2.93	4.59	1.6	2.40	0.4
PC	1.6	2.33	17.9	61.5	7.28	2.96	1.9	3.49	0.8

Table 1.
Chemical composition of the cements.

Cement	Mineral composition (% wt.)			
	C_3S	C_2S	C_3A	C_4AF
N	44.64	23.64	7.94	9.31
SR	53.36	20.81	0.00	13.97
PC	50.63	13.36	14.28	9.01

Table 2.
Mineral composition of the cements.

The content of SO_3 in SR cement is less than 3.5% by weight, which meets the additional requirements for sulfate-resistant cement for general use according to STN EN 197–1 [4]. The values of SO_3 content in N- and PC-cement are higher than in SR-cement. In the case of N-cement it is approaching and in the case of PC-cement it reaches the criterion for the maximum SO_3 content in sulfate-resistant cement up to 3.5% by weight. SR-cement is characterized by the highest proportion of calcium silicates (C_3S and C_2S) and tetra calcium aluminate ferrite (C_4AF) compared to N- and PC-cement and C_3A content of 0.00% by weight. Zero C_3A content in SR-cement meets the additional requirement for C_3A content for sulfate-resistant cement of type SR 0 according to STN EN 197–1 [4]. Many C_3S and C_2S phases in SR-cement are an opportunity for the formation of larger amounts of $Ca(OH)_2$ during hydration, which by its crystalline character markedly affects the formed pore structure and susceptibility to chemical degradation of a cement matrix. N- and PC-cements contain 7.94% wt. and 14.31% wt. C_3A, respectively. Both exceed the requirement for a maximum C_3A content of up to 5% by weight for the type of sulfate-resistant cement CEM I - SR 5 and therefore they cannot be marked as sulfate-resistant types of cement according to the criteria of STN EN 197–1 [4].

The comparison of basic cement properties is reported in **Table 3**. Rheological characteristics of the mortars are presented in **Table 4**. Early- and 28-day strength in basic water curing (BC) is introduced in **Table 5**. N-cement is characterized by the largest specific surface area, normal consistency, flexural and compressive strength values compared to SR- and PC-cement. It is assumed that due to the

Cement	Specific surface area (m^2/kg)	Normal consistency (% wt.)	Initial and final set (min)	Soundness (mm)
N	766.9	35.0	205/270	0.5
SR	354.9	27.2	185/225	1.0
PC	472.4	30.8	265/315	0.0

Table 3.
Specific surface area and the properties of fresh cement mixtures.

Mortar	Consistency (mm)	Volume density (kg/m^3)	Air content (% vol.)
N	151	2236	4.6
SR	186	2205	6.3
PC	142	2239	4.8

Table 4.
Rheological properties of the fresh mortars.

Mortar	Strength (MPa)			
	compressive		flexural	
	2-day	28-day	2-day	28-day
N	37.1	72.2	7.4	9.2
SR	26.2	52.8	4.7	8.4
PC	31.5	58.0	6.2	8.5

Table 5.
Flexural and compressive strength of the mortars.

higher specific surface area of the N-cement, a denser and therefore less permeable microstructure of the mortar is formed during hydration, as a result of which its sulfate resistance can increase in an aggressive environment.

The normal consistency of the cements represents the water content per amount of cement in the cement paste required to achieve the standardized density [18], expressed in percentage. A higher value of the normal consistency of the N-cement means that a higher amount of water is required to achieve the same density of cement paste than with SR- and PC- cement. The cements meet the criterion for the initial setting according to the requirements of STN EN 197–1 [4] over 60 minutes (strength class 42.5 R) in the case of SR- and PC- cement and over 45 minutes (strength class 52.5 N) in the case of N-cement. The final setting time is not specified by the standard. All types of cement meet the criterion for soundness (volume stability) according to the requirements of STN EN 197–1 [4] that has to be below 10 mm.

The mortars are characterized by different consistency, bulk density, and air content. SR-mortar shows the highest consistency (plasticity 186 mm), lower the N- mortar (151 mm), and the densest was PC mortar (142 mm). In the other words SR- mortar needs less mixing water to achieve the same consistency as N- and PC- mortar. The most probable cause is that the missing tricalcium aluminate (C_3A) phase in SR-cement enables lower binding water consumption to the hydrates of calcium aluminate origin (C-A-H hydrated phase). The rich calcium silicate phase alone is not able to absorb so much water at the beginning of hydration and therefore this cement system is more plastic. This experiment did not deal with adjusting the mortars to the same consistency. The mortars were made with the constant water-to-cement ratio 0.5, and therefore all results are from this viewpoint comparable.

The 28-day volume density, as well as dynamic modulus of elasticity of N-, SR- and PC-mortar, are 2250 kg/m^3, 2290 kg/m^3, and 2260 kg/m^3 as well as 45.3 GPa, 41.8 GPa and 43.1 GPa, respectively. The related strength parameters are reported in **Table 5**.

All cements meet the criteria for minimum compressive strength after 2 and 28 days (initial and standard strengths) according to the requirements of STN EN

197–1 [4]. The mortars with cements of strength class 42.5 R (PC- and SR-mortar) must meet the criteria for compressive strength after 2 days more than 20 MPa and after 28 days more than 42.5 MPa, as required by the STN EN 197–1 [4]. The N-mortar made of high-strength NONRIVAL CEM I 52.5 N meets the strength class of 52.5 N achieving compressive strength after 2 days above 20 MPa and after 28 days above 52.5 MPa. Chemical composition of the mortars after 28 days of basic curing (BC) in the water at (20 ± 1) ° C is reported in **Table 6**.

The PC-mortar is characterized by the highest, N-mortar by lower, and SR-mortar by the lowest SO_3 content. This finding is in good agreement with the SO_3 content observed in the types of cement (**Table 1**). Changes in the SO_3 contents are one of the important subjects of monitoring the effect of sulfate attack on the mortars over 5-year exposure time. Basic parameters of the pore structure of mortars after BC are listed in **Table 7**. The highest specific surface area of all open pores, the lowest median radius of all pores within the radii range 1.82 nm to 0.534 mm and micro-pores between 1.82 nm to 5.25 μm points for the presence of the largest share of micro-pores in N-mortar compared to SR- and PC-mortar. This fact is reflected in the formation of a denser, less permeable pore structure of N-mortar also characterized by the lowest total pore volume, total porosity, and the lower permeability coefficient by one order of magnitude compared to SR- and PC-mortar.

The mineral composition of individual cement types together with their fineness generally influences the microstructure formation during hydration, from which depends the condition of the developed pore structure of mortars. The character of the pore structure subsequently determines the permeability of the mortar against the penetration of sulfate solution into the internal structure. The impermeable pore structure is one of the important properties of the cement matrix in terms of environmental resistance.

Explanatory notes to the **Table 7**: specific surface area of all measured pores - SSA, total pore volume – VTP in the measured range of porosimeter 1,82 nm - 0,534 mm, the median radius of micro-pores in the range of 1,82 nm - 5,25 μm, that of macro-pores between 5,25 μm - 0,534 mm and total pores within pore radii of

Mortar	Ign. loss (% wt.)	Content of the component (% wt.)							
		SiO_2	CaO	Al_2O_3	Fe_2O_3	MgO	SO_3	Cl^-	Na_2O eq.
N	7,93	73.08	15.28	0.95	1.14	0.38	0.82	0.03	0.15
SR	8,45	68.99	18.43	0.91	1.80	0.55	0.69	0.02	0.07
PC	7,59	70.99	16.69	1.23	1.14	1.04	0.92	0.02	0.13

Table 6.
28-day chemical composition of the mortars.

Mortar	SSA (m^2/g)	VTP (cm^3/g)	Pore median of			TP (%)	CP (m/s)
			total pores (nm)	micro-pores (nm)	macro-pores (nm)		
N	6.93	0.070	40.10	30.80	9.30	13.70	7.0×10^{-11}
SR	4.02	0.077	67.90	43.21	24.69	16.20	3.0×10^{-10}
PC	4.85	0.079	59.99	42.24	17.75	14.92	2.0×10^{-10}

Table 7.
Pore structure parameters of the mortars after 28-day basic curing.

1,82 nm - 0,534 mm, total porosity TP estimated among pore radii between 1,82 nm - 0,534 mm and calculated coefficient of permeability CP for water valid within the scope of porosimetry measurements.

3.2 Partially accelerated sulfate resistance test

After 28-day BC in the water at (20 ± 1) °C when the mortars reached naturally developed physical-mechanical properties, microstructure, and pore structure, one-half of the mortars were immersed in 5% sodium sulfate solution and the second half of the specimens was still left in the reference water. The partially accelerated test of sulfate resistance of cement mortar is based on long-term, usually two-year, but in this case, even five-year monitoring of a) changes of physical and mechanical properties by non-destructive and destructive testing and b) changes of microstructure and pore structure in 5% sodium sulfate. The obtained results from the sulfate exposure were compared to each other according to the type of cement as well as with those coming from the reference water.

3.2.1 Changes in physical and mechanical properties

Figure 1 shows the changes in dynamic modulus of elasticity (DME) of different mortars (N – NONRIVAL CEM I 52.5 N, SR - sulfate-resistant, and PC – Portland cement) during 5-year exposure in aggressive 5% Na_2SO_4 solution and reference water after 28 days BC. **Figure 2** presents the percentage decrease in the DME of N-, SR- and PC-mortar in the sodium sulfate compared to the reference water.

The N-mortar and SR-mortar show comparable DME changes, while the PC-mortar is subject to the harmful effect of aggressive sulfate attack.

The length changes of mortars during the 28-day BC and 5-year exposure in 5% Na_2SO_4 and water are illustrated in **Figure 3**. The PC-mortar expands significantly during 5 years of exposure in sulfate up to the level of 18.131 mm/m. This expansion gives evidence of the aggressive action of sodium sulfate. Visual observations of the

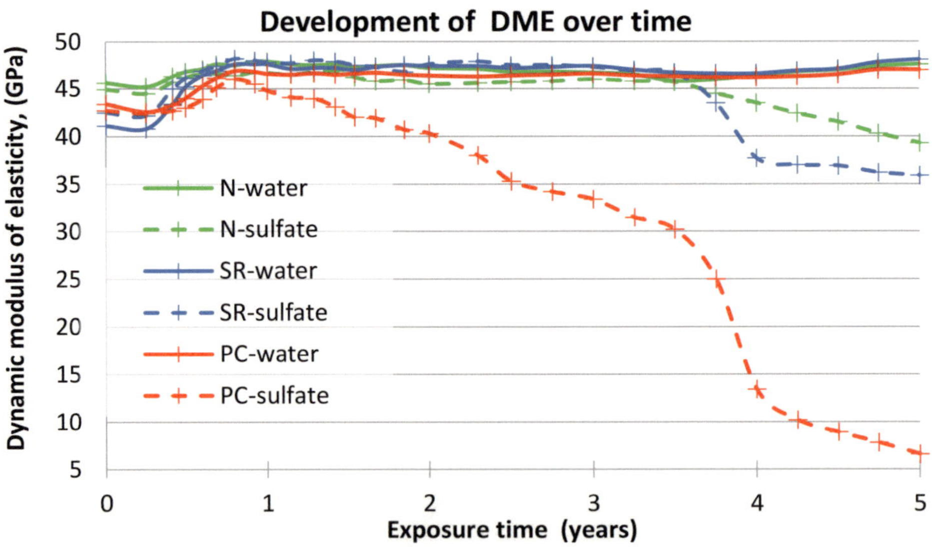

Figure 1.
Changes in dynamic modulus of elasticity of the mortars over time.

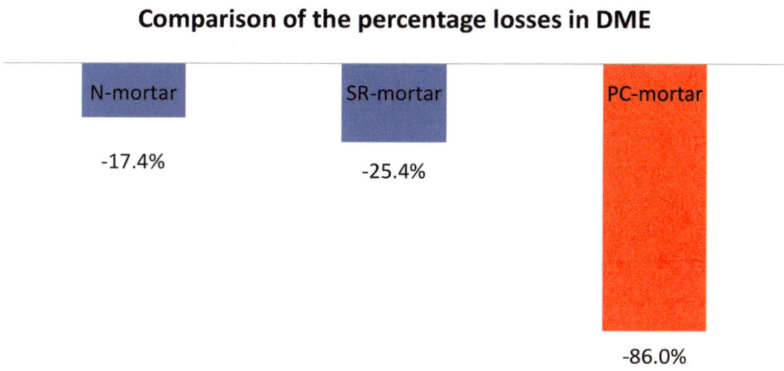

Figure 2.
Percentage loss of dynamic modulus of elasticity of N-, SR- and PC-mortar after 5-year exposure in sodium sulfate compared to the reference water.

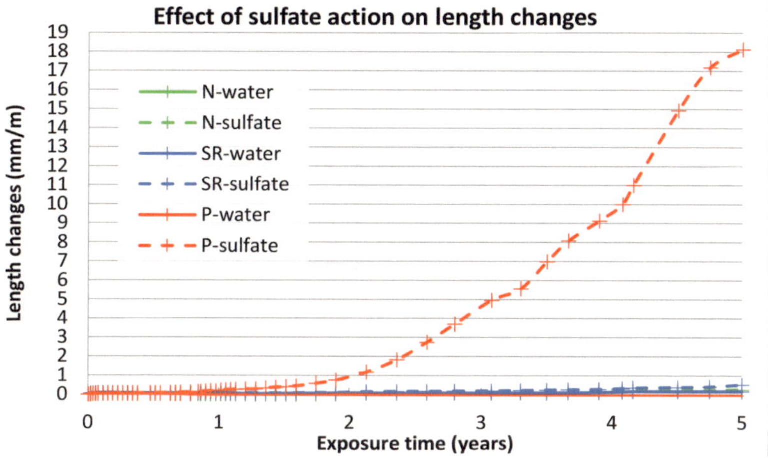

Figure 3.
Length changes of the mortars over time.

PC-mortar show the cracks on the surface of the test prisms, which propagate into the mortar interior over the time. The cracks were mainly observed in the rounding of the prism corners by the loss of the peeled cement matrix. This fact is confirmed by the photo in **Figure 4**.

Figure 5 shows the decrease in compressive strength of all mortars stored in sulfate for 5 years compared to reference water. The largest strength loss is recorded in PC-mortar.

The changes in flexural strength of SR- and N-mortar exposed for 5 years in sulfate are negligible (**Figure 6**). The PC-mortar immersed for the same time in aggressive solution significantly loses the flexural strength to a critical value of 2.8 MPa from 9.6 MPa in water storage. The related loss of flexural strength of PC-mortar is 70.8% wt. (**Figure 6**).

The evaluation of the strength characteristics results in the following partial findings: while N- and SR-mortar show similar sulfate resistance, the apparent strength losses of PC-mortar confirm the well-known evidence that Portland cement is unsuitable for use in a sulfate environment. N-mortar still reaches a sufficiently high strength after 5 years of sulfate attack.

Figure 4.
View on the disturbed surface of PC-mortar exposed for 5 years in 5% sodium sulfate solution.

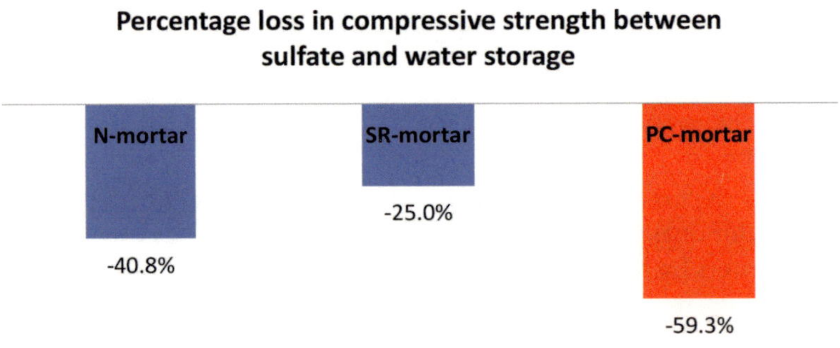

Figure 5.
Percentage loss in compressive strength of N-, SR- and PC-mortar after 5- year exposure in sodium sulfate compared to the reference water.

3.2.2 Changes in microstructure and pore structure

Changes in the physical and mechanical properties of mortars during long-term water and sulfate exposure are a reflection of the formed microstructure and developed pore structure, which mainly depend on the composition and properties of the used types of cement. The mortar's microstructure was studied every year till the end of the experiment by a XRD, thermal and chemical analysis. After 5-year sulfate exposure, the pore structure was identified by MIP and the microstructure observed by the SEM technique. These results serve to elucidate the mechanism of sulfate resistance with special regard to revealing a nature of the sulfate resistance

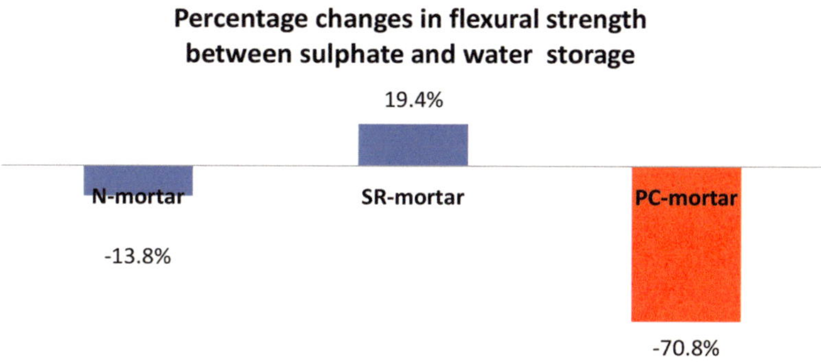

Figure 6.
Percentage changes in flexural strength of N-, SR- and PC mortar after 5-year exposure in sodium sulfate solution compared to the reference water.

of N-mortar. This section presents 5-year results that have made a decisive contribution to clarifying the mechanism of sulfate resistance of NONRIVAL CEM I 52.5 N that is explained in Section 4.

3.2.2.1 X-ray diffraction analysis

XRD analysis determines the qualitative portion of minerals, from which it is not possible to quantify their content, but a comparison of intensity and number of diffractions gives an approximate picture of the mineral content in the mortars. The presence of gypsum (G) and/or ettringite (E) is a decisive indicator of sulfate attack. A comparison of X-ray records of N-mortar after 5 years of exposure in water and 5% Na_2SO_4 is illustrated in **Figure 7**.

Comparison of X-ray records of SR-mortar and PC-mortar after 5 years of exposure in water and 5% Na_2SO_4 is given in **Figures 8** and **9**, respectively.

After 5 years of exposure to sodium sulfate, the PC-mortar shows a high proportion of the formed gypsum (G: $CaSO_4.2H_2O$) as well as also ettringite (E: $3CaO.Al_2O_3.3CaSO_4.32H_2O$) as reaction products of the sulfate attack.

Figure 10 confirms that SR- and N- mortar are characterized by a negligible amount of G and E. Besides these products, every mortar contains portlandite [CH: $Ca(OH)_2$] and quartz coming from a standard sand Q: SiO_2.

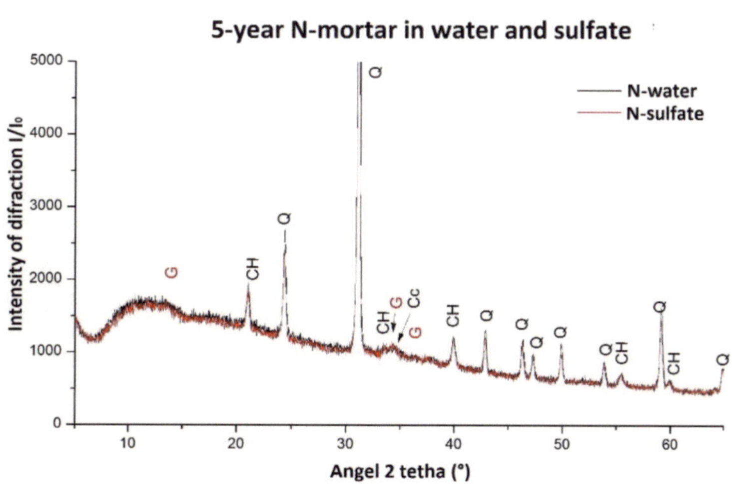

Figure 7.
Mineral composition of 5-year N-mortar in water and sulfate.

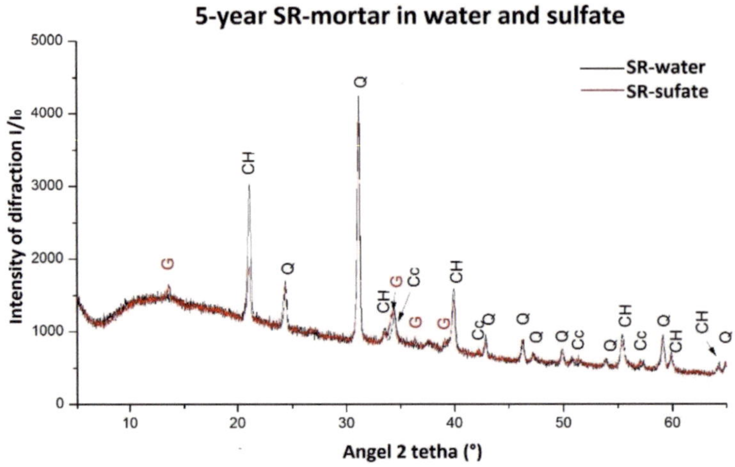

Figure 8.
Mineral composition of 5-year SR-mortar in water and sulfate.

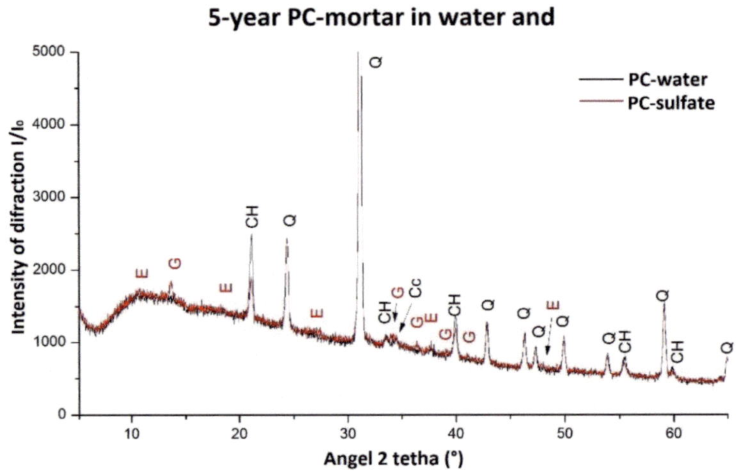

Figure 9.
Mineral composition of 5-year PC-mortar in water and sulfate.

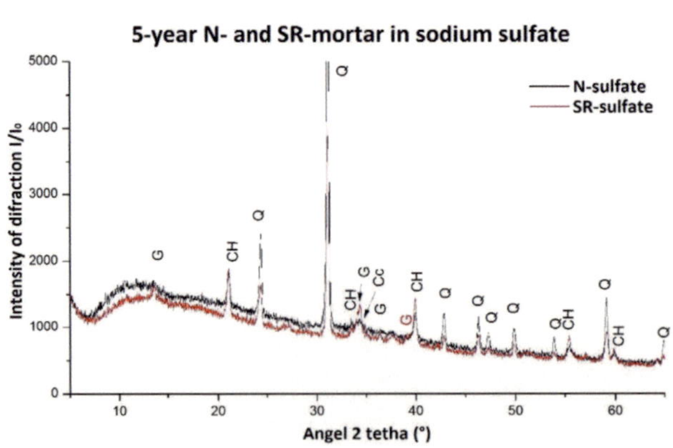

Figure 10.
Comparison of X-diffraction patterns of 5-year N- and SR-mortar from sodium sulfate.

3.2.2.2 Thermal analysis

Thermal analysis (TG-DTA) qualitatively and quantitatively determines the proportion of cement hydration products and products of sulfate attack in the mortars based on the observed mass losses and the relevant endotherms in the respective temperature ranges. The dissociation energy provides information on the incorporation strength of individual releasable components identified by the XRD technique.

The results of the thermal analysis of the 5-year mortars are shown in **Figures 11–13**. The percentage values of present phases and related dissociation energies are reported in **Table 8**. According to **Table 8**, the N-mortar in water contains the lowest proportion of portlandite - $Ca(OH)_2$. All mortars show a decrease in portlandite content after 5 years of exposure to sodium sulfate compared to water. The lowest 0.29% wt. loss is observed in the N-mortar. PC-mortar is characterized by portlandite decrease 3.25% wt. and SR-mortar 4.93% wt. The difference in energy required for the endothermic reaction in the temperature range of 100–200°C in sulfate solution and water can be taken as a measure of gypsum and ettringite incorporation in the mortar's microstructure. The energy value for N-mortar is 3.05 J/mg, for SR-mortar 13.92 J/mg and for PC-mortar 22.83 J/mg. The quantitative representation of G and E as the

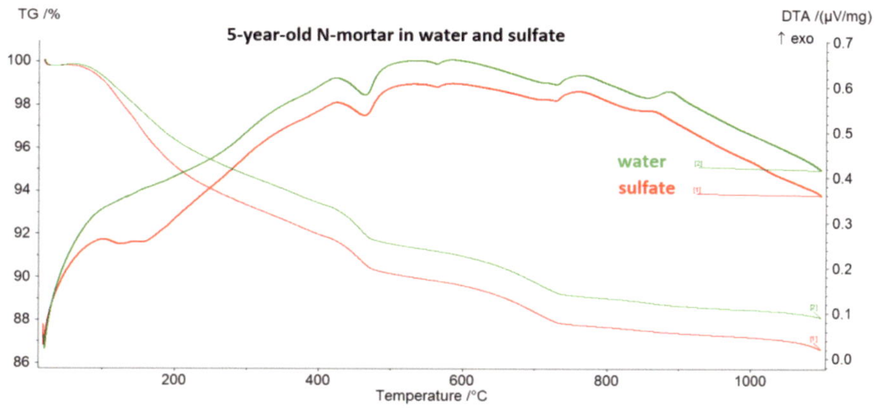

Figure 11.
Comparison of TG-DTA plots of 5-year-old N-mortar in water and sulfate.

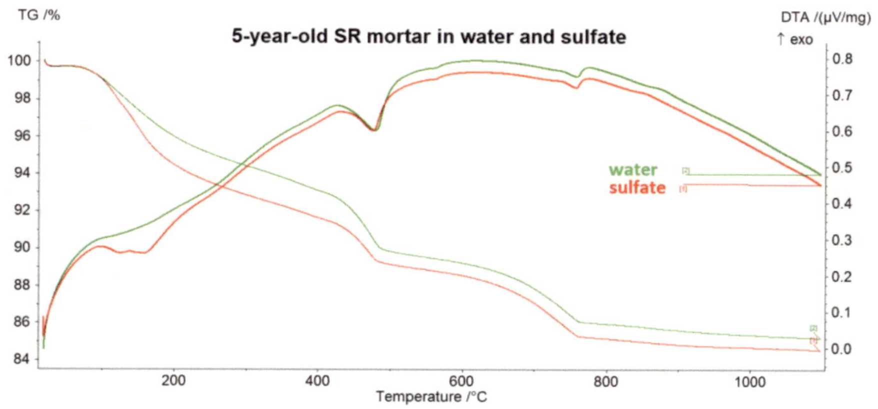

Figure 12.
Comparison of TG-DTA plots of 5-year-old SR-mortar in water and sulfate.

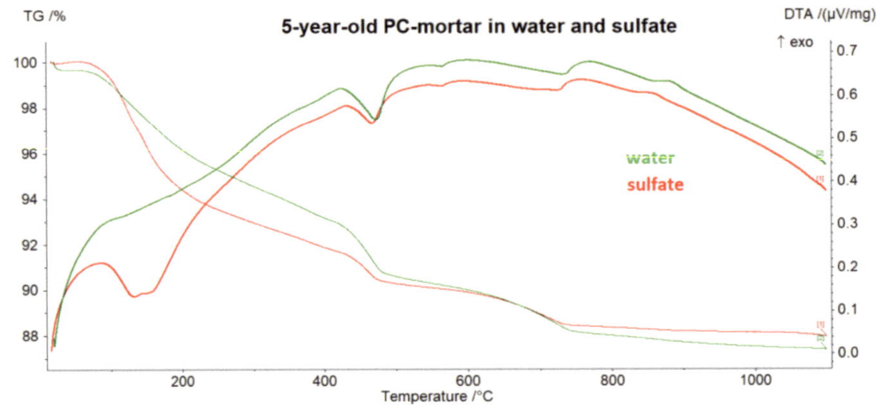

Figure 13.
Comparison of TG-DTA plots of 5-year-old PC-mortar in water and sulfate.

Mortar	Total ignition loss between (20–1100) °C (% wt.)	Content of		Loss in $Ca(OH)_2$ in sulfate (% wt.)	Dissociation energy between (100–200) °C (J/mg)	Energy difference between (100–200) °C (J/mg)
		$Ca(OH)_2$ (% wt.)	$Ca(CO)_3$ (% wt.)			
N-water	11.86	10.03	6.46	0.29	0.00	3.05
N-sulfate	13.35	9.74	6.71		3.05	
SR-water	14.85	15.91	9.14	3.25	0.00	13.92
SR-sulfate	15.44	12.66	9.1		13.92	
PC-water	12.64	13.07	5.98	4.93	0.00	22.83
PC-sulfate	12.06	8.14	4.43		22.83	

Table 8.
Phase composition of the mortars after 5 years of exposure in 5% sodium sulfate solution and reference water at (20 ± 1) °C.

products of sulfate attack is in N-mortar and SR-mortar equally marginal and even the same. On the contrary, PC-mortar shows at the same time the evident presence of gypsum and ettringite.

PC-mortar contains the most quantum of reaction products of the sulfate attack compared to SR- and N-mortar as previously approved by the X-ray analysis. TG-DTA data indicate the equally high sulfate resistance of the N- and SR-mortar.

3.2.2.3 Chemical analysis

The chemical analysis was used to compare the oxide content in mortars with a focus mainly on the SO_3 content. The analysis of the chemical composition does not unambiguously determine the proportion of sulfate attack reaction products but a comparison of the SO_3 content bound in these reaction products gives a picture of the intensity of the acting sulfate aggressiveness. The increase in SO_3 content in the mortar is due to the penetration of sulfate ions from the solution into the internal matrix and the transformation of calcium hydroxide and/or calcium silicate hydrate resp. calcium aluminate hydrate (C-S-H/C-A-H) to $CaSO_4.2H_2O$ and $3CaO.Al_2O_3.3CaSO_4.32H_2O$ [1–3]. A typical symptom of sulfate attack of cement mortar is an increased SO_3 content, which announces the presence of gypsum and ettringite.

This fact is confirmed by the increase of SO_3 in the mortars during 5 years of exposure to 5% Na_2SO_4 (**Figure 14**). The content of SO_3 is in the N-, SR- and PC-mortar after 28-day BC was 0.82%, 0.69% and 0.92% wt., respectively. The 5-year exposure records an increase in the bound SO_3 content to the value of 3.60% wt. in N-mortar, 3.75% wt. in SR-mortar and 6.20% wt. in PC-mortar, while the content of SO_3 in 5-year water storage is 1.54% wt. for N-mortar, 1.31% wt. for SR-mortar and 1.65% wt. for PC-mortar.

The chemical composition of the studied mortars with the attention focused on the typical symptom of sulfate attack - the SO_3 content - shows that N- and SR- mortar after 5-year exposure in 5% Na_2SO_4 are equally characterized by a slight increase in the bound SO_3 content. In contrast, PC-mortar shows at the same time an evident increase of SO_3. Bearing in mind the previous findings of XRD and thermal analysis, chemical analysis points to the fact that the sulfate resistance of the N- and SR-mortar is very similar, even the same, and that both types of cement could be, from this point of view, fully comparable.

3.2.2.4 Pore structure

Basic pore structure parameters after 5 years of exposure to water and sulfate are listed in **Table 9**. The knowledge gained so far suggests that the reaction products (gypsum -G and ettringite -E) are formed during a sulfate attack, which first densifies the pore system. After depletion of the pore storage space by the voluminous G and E reaction products, a loss in the integrity of the formed microstructure starts to occur. In the advanced stage of the sulfate attack when there is observed a loss in mechanical properties and intense expansion, the mortar is characterized by the pore structure coarsening, in particular by the increased porosity. The increased porosity leads to the easier permeability of aggressive sulfate into the internal mortar body. The most evident indicators of these manifestations are the changes in the total pore median radius and total porosity. Changes in basic parameters of the pore structure have therefore a significant impact on the increased permeability.

N-mortar, regardless of the exposure either in water and or the aggressive sulfate, shows in the time horizon of the experiment only a slight pore structure coarsening

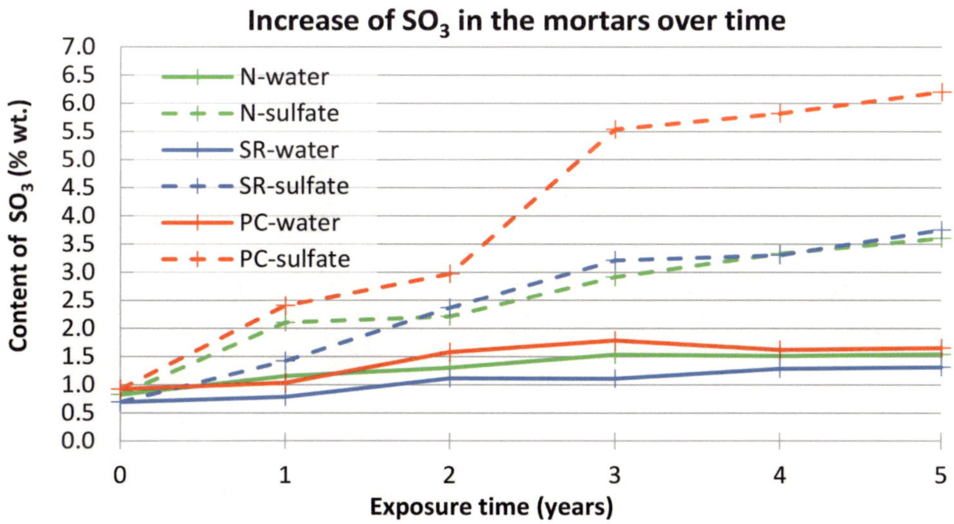

Figure 14.
Changes in SO_3 content in N-, SR- and PC-mortar immersed for 5 years in water and sodium sulfate after 28 days of basic curing in water.

Mortar	SPA (m^2/g)	TPV (cm^3/g)	Pore median radius		TP (%)	CP (m/s)
			of total pores (nm)	of micro-pores (nm)		
N-water	7.10	0.072	33.52	25.76	15.41	2.0×10^{-11}
N-sulfate	5.41	0.070	38.84	26.72	14.32	1.2×10^{-11}
SR-water	6.02	0.071	64.95	35.92	14.95	8.0×10^{-11}
SR-sulfate	5.90	0.065	65.40	28.51	13.49	7.0×10^{-11}
PC-water	6.86	0.072	37.84	25.16	15.25	3.0×10^{-11}
PC-sulfate	5.71	0.083	292.0	2..98	17.03	1.6×10^{-10}

Explanatory notes – the same as for ***Table 7****.*

Table 9.
Comparison of basic parameters of the pore structure of 5-year mortars.

demonstrated by 1) slight decrease in the specific surface area of the measured pores, 2) slight decrease in the total pore volume, 3) a slight increase in the total pore and micro-pore median radii, 4) slight decrease in total porosity in the range of porosimeter measurements and 5) at unchanged permeability at the level of 10^{-11}.

The permeability of the 5-year N-mortar kept in an aggressive sulfate solution is very close to the permeability of the mortar cured for the same time in the water. The reference SR-mortar shows very similar behavior as the N-mortar. The pore structure is not significantly influenced by the sulfate attack. By contrast with it, the PC-mortar is characterized by the typical consequences of sulfate attack proved mainly by an obvious increase in total pore median radius in the aggressive sulfate and the increase in permeability by one order of magnitude to 10^{-10} m/s compared to reference water curing.

3.2.2.5 Scanning electron microscopy

The SEM visually identifies differences in the mortar microstructure between sulfate and water storage. Gypsum crystals have various shapes, most often acicular, prismatic or lenticular, while ettringite crystals are very thin and needle-like. The SEM images of N-mortar are shown in **Figure 15**, while those of PC-mortar in **Figure 16**. The SEM confirms the presence of crystalline $Ca(OH)_2$ in water as

Figure 15.
SEM image of N-mortar after 5-year exposure to water (left) and 5% solution of Na_2SO_4 (right) (magnification 20,000 ×).

Figure 16.
SEM image of PC-mortar after 5-year exposure to water (left) and 5% solution of Na_2SO_4 (right) (magnification 20,000 ×).

well as the presence of gel hydration products of the C-S-H and C-A-H type with minimal occurrence of calcite, also taking into account differences depending on the type of cement. After 5 years of exposure to sodium sulfate, N-mortar records a negligible presence of rod-shaped crystallites (**Figure 15**), while the proportion of these reaction products, which most likely belong to gypsum and ettringite, is dominant in PC-mortar (**Figure 16**).

Summarization of the knowledge according to evaluation of the formed microstructure (XRD, TG-DTA, chemical analysis, SEM) and pore structure (MIP) of the mortars after 5 years of exposure is:

1. the N-mortar made with NONRIVAL CEM I 52.5 N is characterized by the same resistance to the aggressive sodium sulfate in terms of maintaining mechanical properties and structural integrity as the reference SR-mortar made with sulfate-resistant cement of none C_3A content;

2. 2 NONRIVAL CEM I 52.5 N, therefore, declares the same sulfate resistance as the sulfate-resistant CEM I 42.5 R – SR;

3. PC-mortar shows disrupted structural integrity and confirms the well-known fact that ordinary Portland CEM I 42.5 R is not resistant to sulfate aggressiveness.

4. Explanation of the cause of sulfate resistance of NONRIVAL CEM I 52.5 N

Degradation of the hydrated phase of the cement matrix by aggressive sulfate is characterized by the formation of gypsum $CaSO_4 \times 2\,H_2O$ (CSH_2) together with ettringite $3CaO.Al_2O_3.3CaSO_4.32H_2O$ ($C_6AS_3H_{32}$). Gypsum is formed by the reaction of sulfate ions with calcium hydroxide $Ca(OH)_2$ or with calcium silicate hydrate (C-S-H).

$$2\,OH - + SO_4{}^{2} - + Ca^{2+} \rightarrow CSH_2 \text{ or} \tag{1}$$

$$C-S-H + SO_2 \rightarrow CSH_2 \text{ when reaction}\,(1,2)\,\text{takes place in}\,Na_2SO_4\text{ solution} \tag{2}$$

$$2\,Na^+ + SO^{2-} + Ca^{2+} + aq. \rightarrow 2\,Na^+ + CSH_2 + aq. \tag{3}$$

The formed gypsum binds with tricalcium aluminate (C_3A) mainly to ettringite but monosulfate C_4ASH_{12} is also secondarily present.

$$C_3A + 3\,CSH_2 + 26\,H \rightarrow C_6AS_3H_{32} \tag{4}$$

Ettringite formation is accompanied by another minor reaction.

$$C_4ASH_{12} + 2\,CSH_2 + 16\,H \rightarrow C_6AS_3H_{32} \tag{5}$$

The damage mechanism is defined by gypsum and ettringite formation as the reaction products of aggressive sulfate action, being formed as high bulk, voluminous salts that cause destructive expansion of a cement stone. In the reaction 1 and 3, respectively active submicron-sized pozzolan present in NONRIVAL CEM I 52.5 N binds CaO from a supersaturated solution of Ca^{2+} OH^- by the pozzolanic reaction so that the formation of $Ca(OH)_2$ and thus gypsum CSH_2 is extensively eliminated.

Such limited $Ca(OH)_2$ formation due to the pozzolanic reaction of submicron-sized addition subsequently prevents an excessive formation of generated gypsum, required for the ettringite development. The markedly reduced formation of $Ca(OH)_2$ required for the reaction with sulfate ions (see Eq. 1) is the basic condition for suppressing the aggressive effect of sulfate solution on the NONRIVAL CEM I 52.5 N - containing mortar. This is regarded as a new effective reaction mechanism leading in the final effect to the increased sulfate resistance of NONRIVAL CEM I 52.5 N to the same level as that of a sulfate-resistant cement with none of C_3A.

Sulfate-resistant cement (SR) is characterized by blocking gypsum formation and subsequent ettringite due to the absence of C_3A, while NONRIVAL CEM 52.5 N blocks the formation of these reaction products by minimizing the $Ca(OH)_2$ content by the presence of active submicron-based pozzolan. Both alternatives to preventing sulfate aggression mitigate the formation of CSH_2 to a harmless content level but in a different way.

5. Conclusions

Five-year tests of the mortars in 5% sodium sulfate solution show the following key conclusions:

Sulfate resistance of cement NONRIVAL CEM I 52.5 N is the same as sulfate-resistant cement CEM I 42.5 R - SR. The cause lies in the thorough elimination of $Ca(OH)_2$ formation by the active submicron-based pozzolanic addition. The formation of gypsum and ettringite is therefore extensively minimized to harmless content. Reference CEM I 42.5 R does not confirm the resistance to the sulfate solution. For the needs of construction practice, NONRIVAL CEM I 52.5 N represents an equivalent alternative to the use of sulfate-resistant cement in terms of resistance to sulfate aggression.

Other crucial findings coming from the 5-year experiment are:

NONRIVAL CEM I 52.5 N can be advantageously applied in technically demanding structural concrete, in which high strength but at the same time low permeability of the concrete foundation slab is required as a basic condition for ensuring its durability when exposed to aggressive sulfate for a long time. High strength and low penetration permeability are two equally important conditions for achieving a high durability.

Further research should focus on verifying a long-term performance of NONRIVAL CEM I 52.5 N in other aggressive environments.

The recommendation beyond domestic research is:

The development of the unified method, at the best standardized in EN or ASTM, even one common, for the evaluation of concrete's resistance to sulfate attack by an accelerated testing procedure, is required.

Equally urgent scientific task is the most accurate transformation of the accelerated test results to a realistic estimation of concrete service-life when subjected to natural aggressiveness e. g. related to the XA (1–3) exposure classes of STN EN 206 + A1.

Acknowledgements

The financial support of this research project based on the contract related to the utility properties and durability of developed new cement kinds by the Považská cementáreň. a. s. cement plant, Ladce (Slovakia), is greatly appreciated.

Conflict of interest

The authors declare no conflict of interest.

Author details

Michal Bačuvčík[1*], Pavel Martauz[2], Ivan Janotka[1] and Branislav Cvopa[2]

1 Building Testing and Research Institute, Bratislava, Slovakia

2 Považská cementáreň cement plant, Ladce, Slovakia

*Address all correspondence to: bacuvcik@tsus.sk

References

[1] Marchand J. Odler I. Skalmz J.P.: Sulfate attack on concrete. 1st ed. CRC Press: London; 2001. 213 p. ebook ISBN:9780429219337

[2] Collepardi. M. The new concrete. 1st ed. ENCO s.r.l.: Mirano; 2006. 421 p. ISBN 88-901469-4-X

[3] Tang *S. Yao* Y. Andrade C. Li Z: Recent durability studies on the concrete structure. Cement and Concrete Research. 2015. Vol. 78 (Part A). p. 143-154. URL https://www.scipedia.com/public/Tang_et_al_2015b

[4] STN EN 197-1: Cement. Part 1: Composition. specifications and conformity criteria for common cements. Bratislava: Slovak Office of Standards. Metrology and Testing; 2012

[5] Bačuvčík. M. Long-term sulfate resistance of C3A - containing cement NONRIVAL of Ladce provenance. In: Proceedings of the Workshop of Norwegian grant NOVACEM Increasing environmental protection by innovative advanced technologies of cement production of the new generation. 5.-6. May. 2015; Chorvátsky Grob. Slovakia. p 12-13. ISBN 978-80-971912-2-1.

[6] Shi C. Jiménez. F. Palomo A: New cement for the 21st century: The pursuit of an alternative to Portland cement. Cement and Concrete Research. Vol. 41. 2011. p. 750-763. http://dx.doi.org/10.1016/j.cemconres.2011.03.016

[7] Gursel Petek A. Masanet E P. Horvath A. Stadel A: Life-cycle inventory analysis of concrete production: A critical review. Cement and Concrete Composites. 2014. Vol. 51. p. 38-48. DOI: 10.1016/j.cemconcomp.2014. 03.005

[8] Santhanan M. Cohen M D. Olek J: Effect of gypsum formation on the performance of the cement mortar during external sulfate attack. Cement and Concrete Research. 2003. Vol. 33. p. 325-332. DOI: 10.1016/S0008-8846(02)00955-9

[9] Yu C.H. Sun. W. Scrivener K: Mechanism of expansion of mortars immersed in sodium sulfate solutions. Cement and Concrete Research. 2013. Vol. 43. p. 105-111. DOI: 10.1016/j.cemconres.2012.10.001

[10] Müllauer W. Beddoe R E. Heinz. D.: Sulfate attack expansion mechanisms. Cement and Concrete Research. 2013. Vol. 52. p. 208-215. https://doi.org/10.1016/j.cemconres.2013.07.005

[11] Tosun-Felekoglu K: The effect of C3A content on sulfate durability of Portland limestone cement mortars. Construction and Building Materials. 2012. Vol. 36. pp. 437-447. https://doi.org/10.1016/j.conbuildmat. 2012. 04.091

[12] STN EN 206 + A1 Concrete. Specification. performance. production and conformity. Bratislava: Slovak Office of Standards. Metrology and Testing; 2017

[13] STN EN 206/NA Concrete. Specification. performance. production and conformity. Bratislava: Slovak Office of Standards. Metrology and Testing; 2015

[14] Janotka. I.. Martauz. P. Bačuvčík. M. Václavík. V.: Fundamental properties of industrial hybrid cement important for application in concrete. In Pavlo Krivenko editor. Intech Open: London. Open Acess Compressive Strength of Concrete. Edt. by Pavlo Kryvenko. 2020. ISBN: 978-1-78985-568-5; p. 1-25.

[15] Saleh H.M. Tawfik, M. A. Bayoumi T.A.: Chemical stability of seven years aged cement–PET composite waste form containing radioactive borate waste simulates.

Journal of Nuclear Materials. 2011. Vol. 411. 2011. p.185-92. DOI: 10.1016/j.jnucmat.2011.01.126

[16] Eskander B. Bayoumi T.A. Saleh H. M.: Performance of aged cement–polymer composite immobilizing borate waste simulates during flooding scenarios. Journal of Nuclear Materials. 2012. Vol. 420. p. 175-181. DOI: 10.1016/j.jnucmat.2011.09.029

[17] Zhutovsky, S. Hooton, R. D.: Accelerated testing of cementitious materials for resistance to physical sulfate attack. Construction and Building Materials. 2017. Vol. 145. p. 98-106. DOI:10.1016/j.conbuildmat.2017.03.239

[18] Gu, L. Visintin, P. Bennet, T.: Evaluation of accelerated degradation test methods for cementitious composites subject to sulfuric acid attack application to conventional and alkali/activated concretes. Cement and Concrete Composites. 2018. Vol. 87. p. 187-204. DOI: 10.1016/j.cemcon comp. 2017.12.015

[19] STN EN 196-2 Methods of testing cement. Part 2: Chemical analysis of cement. Bratislava: Slovak Office of Standards. Metrology and Testing; 2013

[20] STN EN 196-3 Methods of testing cement. Part 3: Determination of setting time and soundness. Bratislava: Slovak Office of Standards. Metrology and Testing; 2020

[21] STN EN 196-1 Methods of testing cement. Part 1: Determination of strength. Bratislava: Slovak Office of Standards. Metrology and Testing; 2019

[22] STN EN 1015-3 Methods of test for mortar for masonry. Part 3: Determination of consistence of fresh mortar (by flow table). Bratislava: Slovak Office of Standards. Metrology and Testing; 2000

[23] STN EN 1015-6 Methods of test for mortar for masonry. Part 6: Determination of bulk density of fresh mortar. Bratislava: Slovak Office of Standards. Metrology and Testing; 2000

[24] STN EN 1015-7 Methods of test for mortar for masonry. Part 7: Determination of air content of fresh mortar. Bratislava: Slovak Office of Standards. Metrology and Testing; 2000

[25] STN 72 2453 Testing of volume stability of mortar. Bratislava: Slovak Office of Standards. Metrology and Testing; 1968

[26] STN 73 1371 Method of ultrasonic pulse testing of concrete. Bratislava: Slovak Office of Standards. Metrology and Testing; 1981

Chapter 4

Impact of Nanosilica in Ordinary Portland Cement over Its Durability and Properties

Gude Reddy Babu, Pala Gireesh Kumar, Nelluru Venkata Ramana and Bhumireddy Madhusudana Reddy

Abstract

The present examination illustrates the impact on the hardened and fresh cement mortar and cement with the inclusion of nanosilica of size 40 nm in various environmental conditions (UltraTech, India). It is quite notified that an elevation in compressive strength as well as flexural strength along with an improvisation in the performance and life span of cement mortar. The samples of M5 grade blended with a ninety percentage of concrete and remaining with nanosilica was identified to have a finer working elevation in as well as in standards when collated with the conventional cement mortar. The corollary of hardened and fresh cement, strength parameters were looked upon with the aid of XRD (X-ray Diffraction). Also, the SEM (Scanning Electron Microscope) test holds a predominant role in analysis.

Keywords: OPC, Strength, HCl, $MgSO_4$, XRD, nanosilica and SEM

1. Introduction

There is no other substitute for concrete that can be supplemented with an alternative because of its intrinsic qualities like get into any shape, quality and ability to consume locally available fine and coarse materials along with its strength, resistance to fire, with little support [1]. The use of concrete and its consumption is on par with wans and the activity of construction is set to have been emitting 8% of CO_2 [2]. Even though the substitute for concrete is not identified the area of development becomes the centre stage for identifying substitute materials. The contribution of nano technology has made big strides and effect on various aspects of science. Nano silica is preferably utilized in numerous approaches and is easily available as well.

The research and investigation in nano particles and their utilization in cement and mortar invariably becomes popular [3]. The properties of cement in both hardened as well as in fresh phase are majorly laid focus on. The density of concrete and cement can be improvised with the inclusion of nano as well as micro NS which have the tendency of acting as a material of filler, which results in up gradation of its quality [4–17]. Nano silica with separated molecule impact got an extraordinary pozzolonic property than other pozzolonic materials on comparison because of the nano size of particles and greater silica contents, accordingly both impacts are important in making concrete [4, 6]. Expansion of Nano silica to solidify concrete and cement mortar then resulted in improvement of sharp and long haul considers and over utilisation of nano silica results in decrease in quality. Nano silica greatly improved qualities of cement and concrete in fresh and solidified stage and in durability in different environmental conditions [10–11, 18–19]. The nano particles are centered and a considerable research is carried out.

Therefore, the present research includes the impact and durability of 40 nm silica particles in cement mortar, durability existence under different environments both physical and chemical are taken up for examination [20–23]. For acquiring an accurate output, instruments of Scanning Electron Microscope (SEM) and X-ray diffraction (XRD) were resorted to. In this research, cement to sand proposition was 1: 3 and water to cement (concrete plus NS) proposition was set at 0.4.

2. Procedure

In the set forth work area, the proportion of 1:3 is considered for cement-sand and 0.4 for water-cement. Also, nanosilica is incorporated along with cement starting from zero percentage and with every sample, an increase of 2% is followed until an overall 14% is achieved. Every sample is treated individually with starting with M0 and concluding with M7.

2.1 Blending of nano silica with cement

Sticking to the particulars of Jo B W [24], the blending of NS with cement mortar is done for a minute maintaining 285 rpm. Prior to the blending process, the water is treated with NS and this was put into the rotational blender which runs at 140 rpm. This was done for 30 seconds so as to solidify the mixture and addition of a fine total was done. Adding to the process, a super plasticizer was annexed to the mix when the blender was running at a speed of 285 rpm. Now, a resting phase of 90 seconds is allocated to the blender and it resumes working for a minute with the same speed before rest. Finally, the mix was set out in a hardened figure.

2.2 Hardening or setting interval (setting time)

IS of code 5513-1976 was used for detecting the time taken for setting.

2.3 Strength determination (compressive strength)

The part 6 of 4031-1998 was considered to find the compressive quality of mortar.

2.4 Strength determination (flexural strength)

Determining the flexural standards using the BS section 188 of 1881 of the year 1983 was done.

2.5 Durability aspect

AR graded magnesium sulphate and hydrochloric acid, which are synthetic in nature, were used in the research. The ACI of 318-99 were followed and the compound grouping was done. An exemplar of the fixations projects over a span of 120 days for 1%, 240 days for 2%, 360 days for 3% and 480 days for 4%, provided the recharge of the fixations was for each 4-month tenure.

2.6 Temperature assessment

The blended samples of M0 as well as M5 were tested upon for the temperature check at 900 °C in a suppress heater. Also, the blended sample out-righting 28 days were dried for a span of 6 hours at normal temperature and then they were transferred to a muffle furnace. An altering is done with temperatures where each temperature has got three samples assessed for a period of two and half hours. The altering temperatures start from 100 °C and every alteration has an add-on of 100 °C up to 600 °C. Finally, the samples were put out to cool down at a temperature of 30 °C i.e., room temperature (**Tables 1–5**).

Oxides	Percentage
Calcium oxide	62.5
Silicon dioxide	20.5
Aluminium oxide	6.1
Ferrous oxide	3.1
Magnesium oxide	1.6
Sulphur trioxide	1.5
Oxides of potassium and sodium	1

Table 1.
Constitutes of cement.

Attributes	Standard values (IS 12269-1987)	Obtained values
Compressive Strength (days) 3 7 28	27 MPa (Min) 37 MPa (Min) 53 MPa (Min)	35.91 MPa 46.72 MPa 62.90 MPa
Relative Density	—	3.10
Setting Span Initial Final	30 min 600 min	155 min 267 min
Soundness (Le- Chatlier Expansion)	10 mm (Max)	0.85 mm
Specific Surface Area (m^2/Kg)	226	314

Table 2.
Cement attributes.

Ennore sand grading	Value (%)
Passing through Sieve of 2 mm	100
Retained over Sieve of 90 m	100
1 mm < Size of the particle	33.33
500 m < 1 mm > Size of the particle	33.33
500 m > Size of the particle	33.33

Table 3.
Ennore sand attributes.

Attributes	Obtained values
Relative density	1.33
Surface area	50.5
Particle size nm	40
SiO_2	99.7%
Loss on Ignition	0.3%

Table 4.
Attributes of nanosilica.

Attributes	Standard values (IS: 456 – 2000)	Values of distilled water	Values of drinkable water
power of hydrogen, ph	6.5- 8.5	7.2	7.5
Total Dissolved Solids (mg/L)	2000	4.6	11.0
Organic Solids (mg/L)	200	1.1	6.0
Inorganic Solids (mg/L)	3000	6.2	17.0
Alkalinity (mg/L)	250	3.6	11.0
Acidity (mg/L)	50	0	5.0
Sulphates (mg/L)	400	3.2	12.0
Chlorides	500(RCC)	1.5	9.0

Table 5.
Attributes of water.

3. Deliberation of verdict

3.1 Hardening or setting interval (setting time)

Figure 1 reveals the corollary on setting time due to NS and uncovers the fact of lowering the hardening time with increment in NS quantity. The initial and final setting time for M0 was 155 min and 266 min, M1 was 153 min and 263 min, M2 was 148 min and 260 min, M3 was 143 min and 255 min, M4 was 137 min and 245 min, M5 was 131 min and 240 min, M6 was 126 min and 234 min, M7 was 123 min and 227 min. The cycle was advanced because of NS has a vast surface region; subsequently, hydration measure turns out to be quick.

3.2 Analysis of strength (compression)

A rise in the strength upon inclusion of nanosilica up to 10% was portrayed in **Figure 2**. With further addition of NS, decrement in strength can be observed from the diagram. The finest of all samples is the M5 sample comprised of 90% cement and 10% NS. For 3 days, the increment strength observed was 17.8 MPa, 7 days resulted 21.58 MPa, 28 days directed a strength of 21.18 MPa, 90 days showcased 21.50 MPa, 180 days witnessed 21.32 MPa and finally, 365 days exhibited a strength of 20.66 MPa. From the analysis, it is truely clear that M6 blended with 88% of cement and the remaining with NS has seen a decrease in compressive strength whereas M5 figured the best strength notable.

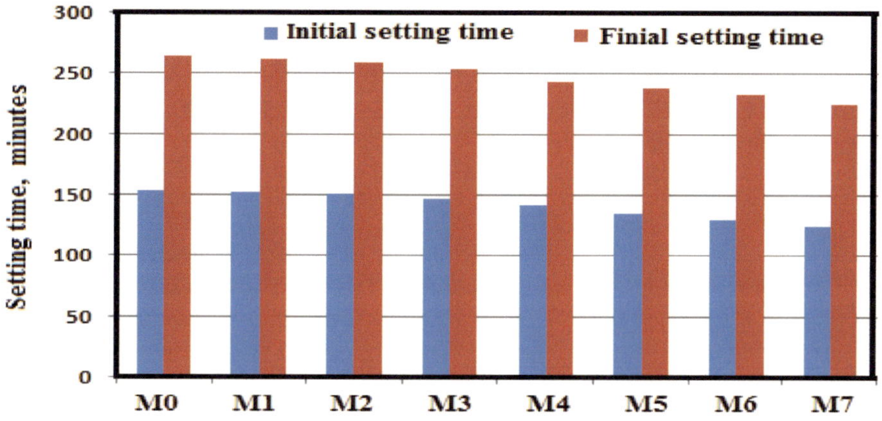

Figure 1.
Efficacy of nanosilica over setting span.

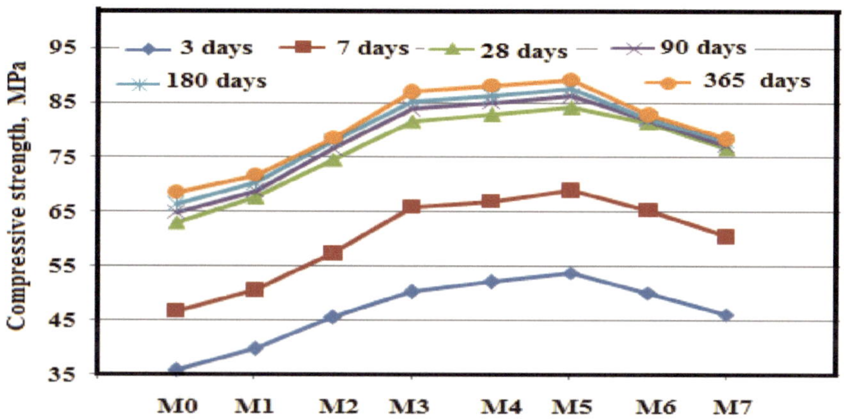

Figure 2.
Efficacy of nanosilica over strength attribute (compression).

3.2.1 Manoeuvre of XRD

The analysis was done for a couple of samples having 0% NS and 10% NS i.e., for M0 and M5 by X-ray diffraction, XRD. The samples were of OPC type and were hydrated for a span of 28 days. **Figure 3** shows that hydroxide of calcium showed up at 18° whereas the calcium silicate hydrate showed up at 26° individually for both the samples. Concentration and strength of hydroxide of calcium was predominantly high and that of silicate hydroxides of calcium was less when compared to OPC of 90% and nano silica of 10% test. The variation of strength for the sample of M0 and the sample of M5 was due to the reaction of oxides of silica with hydroxides of calcium (finished results of cement hydrate).

3.2.2 Manoeuvre of scanning electron microscope

Figures 4 and **5** demonstrate the results for the sample M0 and also for the sample M5 which were hydrated for 28 days. The diagram of Scanning Electron Microscope (SEM) portrays the hydroxides of calcium's needle like structure along with the silicate hydroxide of calcium which was spread widely. The calcium hydroxide's long needle like structures in **Figure 4** can be collated with **Figure 5**.

Also, a high in silicate hydroxide of calcium's content can be seen in **Figure 5** when compared to **Figure 4**. The major difference in the structure of hydroxides of calcium and quantity of silicate hydroxide of calcium was due to the presence of oxides of silica in NS. These react with hydroxides of calcium in the cement hydrate and thereby, causing the differences.

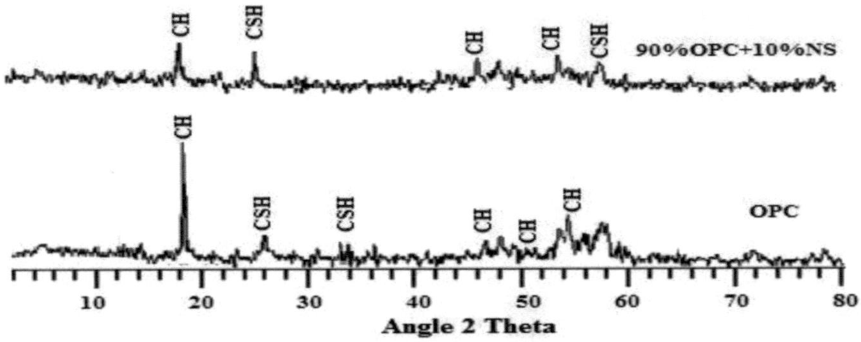

Figure 3.
Depiction of M0 & M5 samples XRD (hydrated for 28 days).

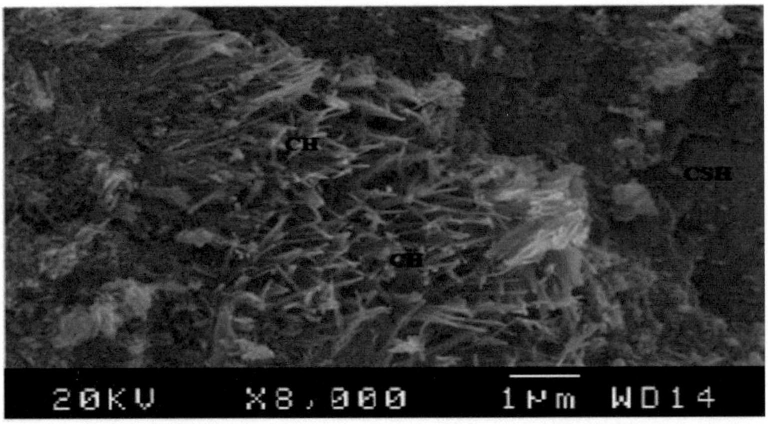

Figure 4.
Depiction of M0's SEM diagram (28 days of hydration).

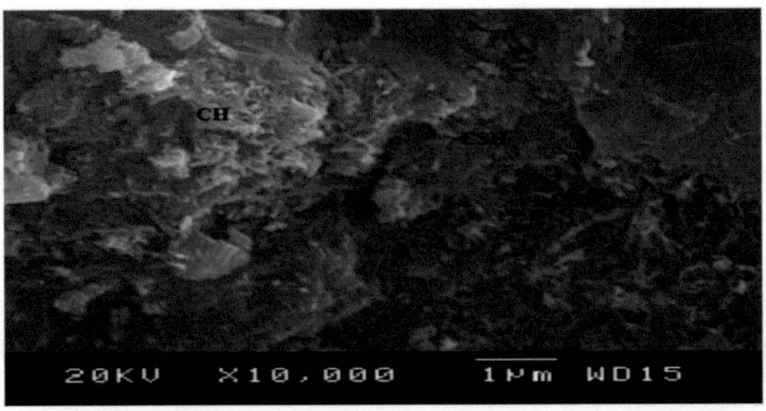

Figure 5.
Depiction of M5's SEM diagram (28 days of hydration).

3.3 Analysis of flexural strength

The NS impact on flexural strength can be identified from the **Figure 6**. It was evident that cement with less than 10% NS exhibited expand in flexural strength by 10% whereas cement with NS in more than 10% showcased decline in the strength. The samples were identified upon M5 blend with a higher expansion rate with time. It was seen that 2 MPa was recorded for 3 days, 2.2 MPa for 7 days, 2.5 MPa for 28 days, 2.4 MPa for 90 days, 2.4 MPa for 180 days and finally 2.6 MPa for 365 days. From the analysis, we can notice that samples has shown a decrease in strength with NS percentage more than 10 but the value was higher when compared with the strength of sample with zero percentage of NS. The major concerns that led the expansion of the strength characteristics were the size of the particles and content of oxides of silica. An extra quantity of silicate hydroxide of calcium was witnessed due to the action of hydroxides of calcium and oxides of silica. Furthering this, the NS acts as a framework's filler material of cement mortar.

3.4 Probes on durability aspect

3.4.1 Corollary of magnesium sulphate

Figures 7 and **8** render the effect of Magnesium Sulphate ($MgSO_4$) over compressive strength of the samples. If one observes in **Figure 7**, by and large, cement mortar strength quality increased irrespective of age, the equivalent can be observed in M0 examples of no $MgSO_4$ fixation., Compressive strength qualities decreased as for time and focus with less than 4% fixation, and 4% centralization of $MgSO_4$ has resulted in the greatest quality decrease.

M5 blend examples equivalents the above and were provided in **Figure 8**. Concerning the concentration and age of $MgSO_4$, M0 blend examples get altogether result in quality decrease compared to M5 blend examples Also the M0's compressive strength for 4% was noted to be 57.30 MPa for 120 days, 53.55 MPa for 240 days, 48.50 MPa for 260 days and finally 44.16 MPa for 480 days. In the similar context, the strength values of M5 were 78.26 MPa for 120 days, 74.1 MPa for 240 days, 67.0 MPa for 260 days and finally 64.0 MPa for 480 days. The contrast of strength characteristic between M0 and M5 was 21, 20.4, 20.3 and 19.7 MPa respectively.

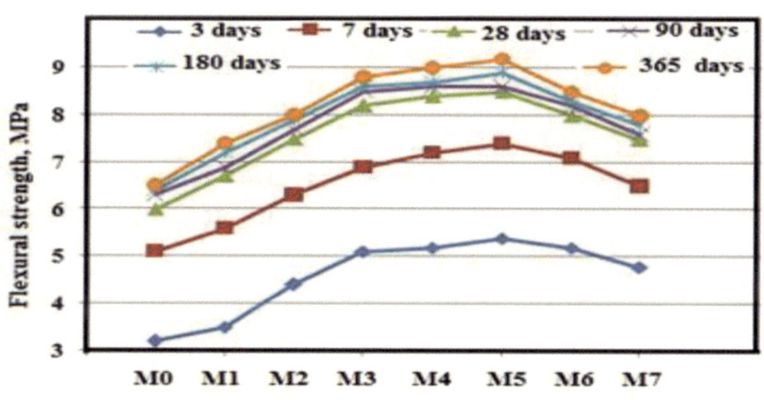

Figure 6.
Efficacy of nanosilica upon strength (Flexural).

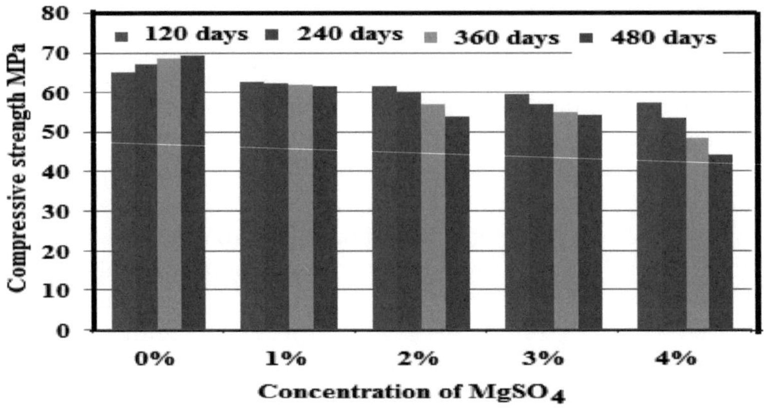

Figure 7.
Corollary of $MgSO_4$ over Sample Mo.

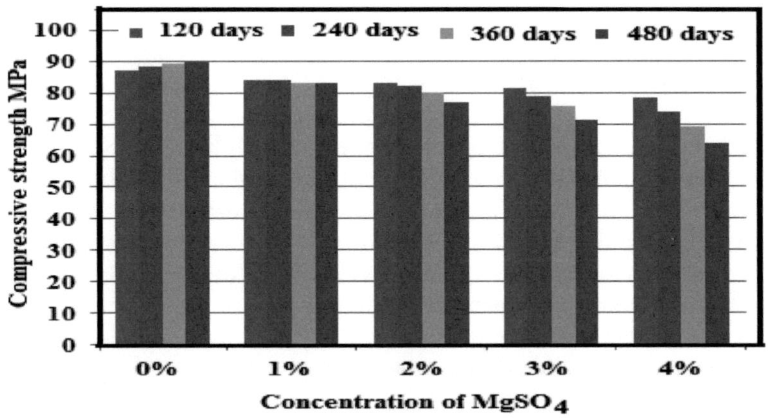

Figure 8.
Corollary of $MgSO_4$ over Sample M5.

3.4.2 Manoeuvre of X-ray diffraction

The blended samples of OPC's M0 and M5 were hydrated for 28 days and the samples were kept in water with 4% of $MgSO_4$ for a span of 360 days. The results of the sample M0 and the sample M5 were represented as charts in the **Figure 9** where magnesium silicate hydrate was presented at 48°; at 18°, the hydroxide of calcium turned up and at 33°, magnesium's hydroxide was presented and finally the silica hydroxide of calcium was shown at 26°. It was noted that the hydroxide of calcium exhibited a high force of ascent and, additionally, the lower power ascent of silicate hydroxide of calcium of M0 diversified from M5. The contrast happened because of nanosilica quality. Also, hydroxide and silicate hydrate of magnesium were directed because of the reaction of $MgSO_4$ with hydroxide and silicate hydroxide of calcium.. The quality of silicate hydrate of magnesium declined due to the non-binding nature, but M5 exhibits a finer standard when collated with M0 due to the inclusion of nanosilica.

3.4.3 Manoeuvre of scanning electron microscope

The blended samples of OPC's M0 and M5 were hydrated for 28 days and the samples were kept in water with 4% of $MgSO_4$ for a span of 360 days. The results

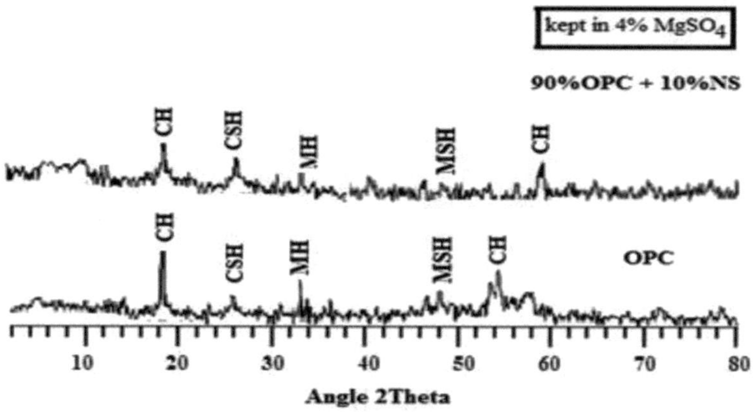

Figure 9.
Depicting M0 & M5 samples XRD (Sample placed for 360 days in water with 4% $MgSO_4$).

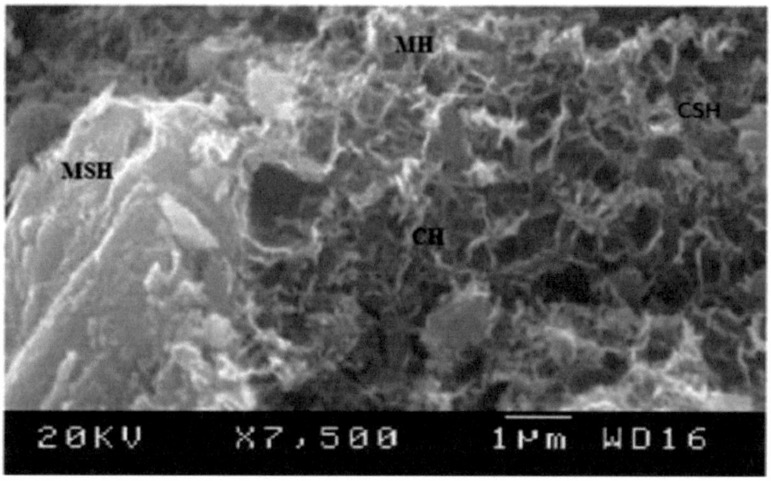

Figure 10.
Depicting M0's SEM diagram (Sample placed for 360 days in water with 4% $MgSO_4$).

were represented as Scanning Electron Microscope diagrams in the **Figures 9** and **10** where hydroxide of calcium, silicate hydroxide of calcium, hydroxide of magnesium and silicate hydrate of magnesium were presented. Calcium hydroxide's needle structure was indicated in the **Figure 10** and further, it was made in contrast with the **Figure 11** whereas the content of calcium silicate hydroxide is high when compared to **Figure 10**. In further addition, magnesium's silicate hydrate and hydroxide witnessed a fine and more elegant outlook when collated with **Figure 11**. The inclusion of NS shows diversified nature of OPC. Also, placing the sample in water mixed with $MgSO_4$ adds up to the work.

3.5 Corollary of hydrochloric acid

Figures 12 and **13** render the effect of Hydrochloric Acid (HCl) over compressive strength of the samples. By and large, concrete mortar quality expanded irrespective of age, the equivalent can be seen at sample M0 which was placed in zero HCl fixations. It was also seen that a decline in compressive quality with concentration and time in **Figure 12** when M0 blend examples kept in less than 4% convergence of HCl and examples were given for HCl centralization with 4% has the greatest quality decrease.

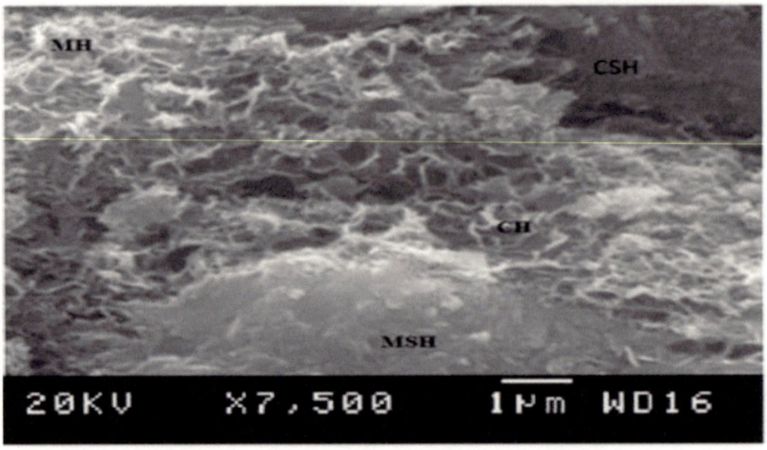

Figure 11.
Depicting M5's SEM diagram (Sample placed for 360 days in water with 4% $MgSO_4$).

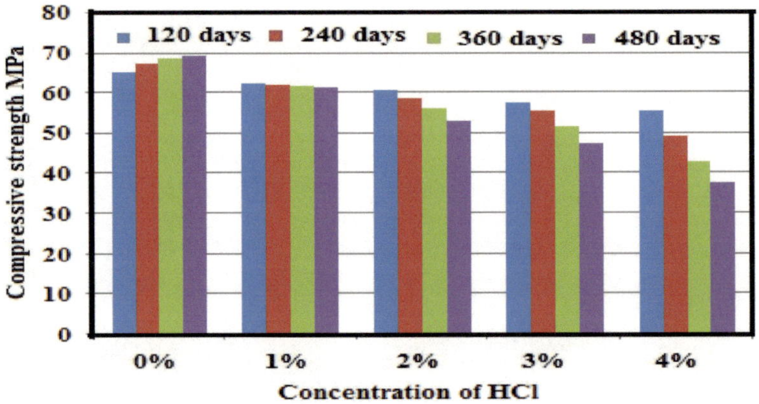

Figure 12.
HCl Corollary over Sample M0.

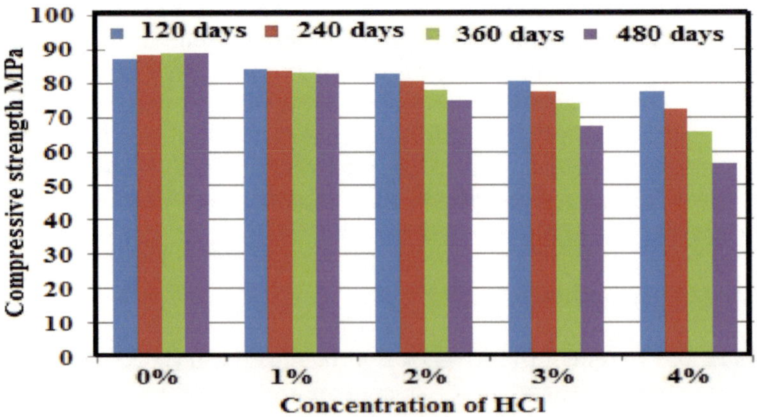

Figure 13.
HCl Corollary over Sample M5.

The equivalent to above can likewise be seen in M5 blend examples which appear in **Figure 13**. Yet, M0 blend examples get noteworthy quality decrease contrasted with that of M5 blend examples. M0's and M5's compressive quality for a focus of 4% was is 55.35 MPa and 77.33 MPa for 120 days, 49.00 MPa and 72.30 MPa for 240 days, 42.60 MPa and 65.55 MPa for 260 days and 37.31 MPa and 56.30 MPa for 480 days. The differentiation in strength at 4% fixation for 120 days was 21.3 MPa, 240 days was 23.25 MPa, 260 days was 22.82 MPa and for 480 days was 18.96 MPa.

3.5.1 Manoeuvre of X-ray diffraction

The blended samples of OPC's M0 and M5 were hydrated for 28 days and the samples were kept in water with 4% of HCl for a span of 360 days. The results were represented in the **Figure 14** where hydroxide of calcium, silicate hydroxide of calcium, chlorides of calcium, Friedel's mixes of salt of M0 sample and M5 sample were presented individually at 18°, 26°, 33°, 37° and finally at 48.5°. It was notified that the ascent power of calcium hydroxide is high whereas the force ascent of silicate hydroxide of calcium is more for M5 than M0. The observed variation in ascent of M0 sample and M5 sample was majorly due to the expansion taking place in cement mortar due to the inclusion of NS. Adding to it, the reaction of hydroxides of calcium with HCl and chlorides of calcium with C_3A frames the chlorides of calcium and Friedel's mixes of salt. Silicate hydroxide of calcium got destabilized due to its reaction with HCl resulting a decline in the quality of strength but the performance of M5 was way too good when collated to that of M0 because of essence of NS in cement mortar.

3.5.2 Manoeuvre of scanning electron microscope

The blended samples of OPC's M0 and M5 were hydrated for 28 days and the samples were kept in water with 4% of HCl for a span of 360 days. The results were represented as Scanning Electron Microscope diagrams in the **Figures 15** and **16** where hydroxides of calcium, silicate hydroxides of calcium, chlorides of calcium were shown. A significant improvement in case of above mentioned compounds can be noted in the **Figure 15** when contrasted with the **Figure 16**. The diagrams of Scanning Electron Microscope test uphold the investigation of X-ray Diffraction.

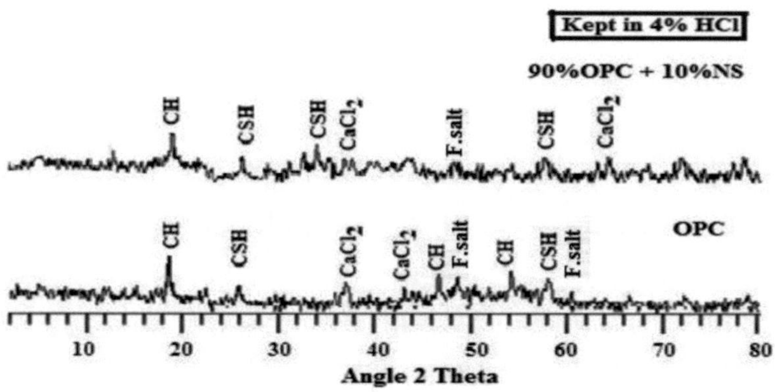

Figure 14.
Depicting M0 & M5 samples XRD (Sample placed for 360 days in water with 4% HCl).

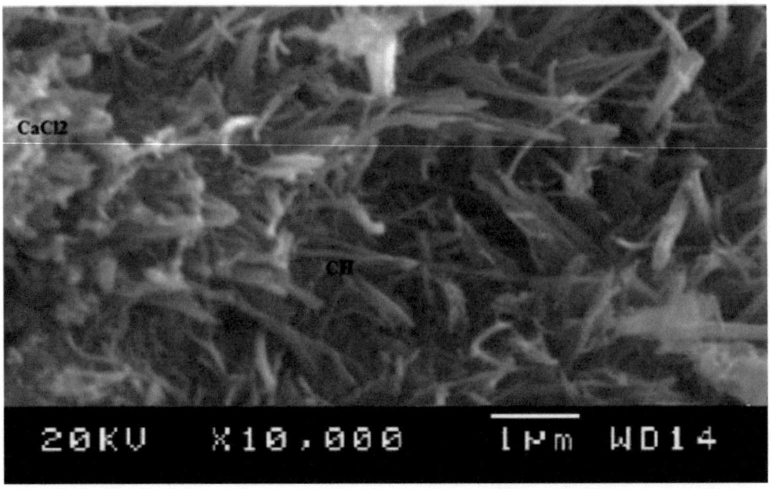

Figure 15.
Depicting M0's SEM diagram (Sample placed for 360 days in water with 4% HCl).

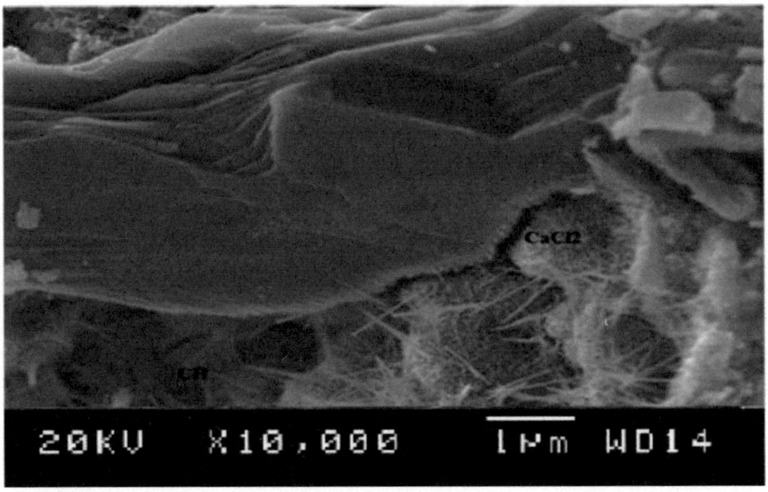

Figure 16.
Depicting M5's SEM diagram (Sample placed for 360 days in water with 4% HCl).

3.6 Assessment of temperature corollary

Temperature's corollary on the blend samples of M0 and M5 were shown in the **Figure 17**. It was evidently clear that an increase in temperature of M0 and M5 samples results the compressive strength to lower. The strength values noted for M0 sample and M5 sample at various temperature conditions are 62.93 MPa and 84.12 MPa at 27 °C, 51.90 MPa and 73 MPa at 400 °C, 28.30 MPa and 48 MPa at 600 °C, 15.40 MPa and 28.41 MPa at 800 °C. The M0 sample and M5 sample exhibits decrease in compressive strength quality because of between layer water, losing free water and synthetically reinforced water. Furthermore warm extension of cement mortar and was unique. The quality decreases in case of compressive strength for sample of M0 and sample of M5 was critical over a temperature of 400 °C. But, more compressive quality was exhibited by M0 blend examples than M5 blend examples.

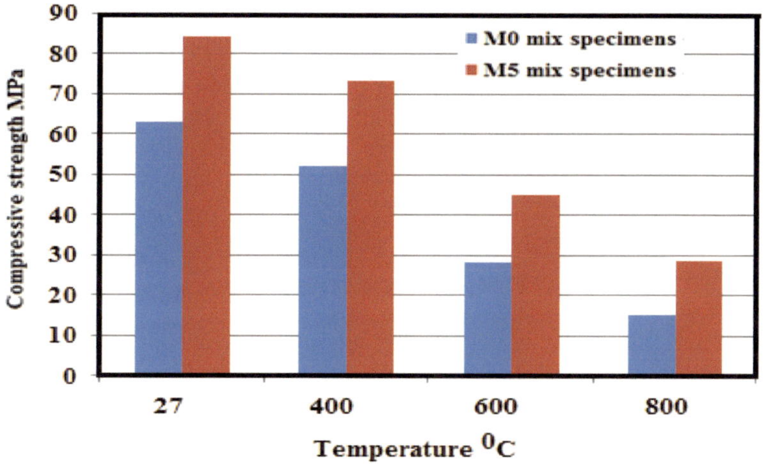

Figure 17.
Corollary of temperature.

4. Conclusion

The strength qualities (compression and flexural) expanded by and large due to the NS of 40 nm in case of M5 blend. Cement's extraordinary pozzolonic property and its property as a good material for filling of NS instigated the rise in strength with cement and brilliant filling material.

NS has quickened setting measure because of its enormous surface area.

M5 mix examples have given much better execution in solidness in different environmental conditions than that of M0 blend examples.

Author details

Gude Reddy Babu[1*], Pala Gireesh Kumar[2], Nelluru Venkata Ramana[3] and Bhumireddy Madhusudana Reddy[4]

1 Department of Civil Engineering, Gudlavalleru Engineering College, Gudlavalleru, Andhra Pradesh, India

2 Department of Civil Engineering, Shri Vishnu Engineering College for Women (A), Bhimavaram, Andhra Pradesh, India

3 Construction Technology Department, V.T.U P.G. Centre, Kalaburagi, Karnataka, India

4 Department of Civil Engineering, Sri Venkateswara University College of Engineering, Tirupati, Andhra Pradesh, India

*Address all correspondence to: greddybabu1966@gmail.com

References

[1] P.K. Mehta, H. Meryman, Tools for reducing carbon emissions due to cement consumption, Structure magazine. (2009) pp. 11-15.

[2] J. G. J. Olivier, J.–M. Greet, J. A. H. W. Peters, Trends in global CO2 emissions 2012 report, PBL Netherlands. Environmental Assessment Agency (2012) pp. 17.

[3] I. Ali, New generation adsorbents for water treatment, Chem. Rev. 112 (2012) 5073-5091.

[4] Y. Qing, Z. Zenan, K. Deyu, C. Rongshen, Influence of nano-SiO2 addition on properties of hardened cement paste as compared with silica fume, Constr. Build. Mater. 21 (2007) 539-545.

[5] A. Sadrmomtazi, A. Fasihi, F. Balalaei, A.K. Haghi, Investigation of mechanical and physical properties of mortars containing silica fume and nano-SiO2, in: Third International Conference on Concrete and Development, Tehran, Iran, 27-29 April, 2009, pp. 1153-1161.

[6] H. Li, H. G. Xiao, J. Yuan, J. P. Ou, Microstructure of cement mortar with nanoparticles, Compos. Part B 35 (2004) 185-189.

[7] H. Li, H. G. Xiao, J. Yuan, J. P. Ou, A study on mechanical and pressure-sensitive properties of cement mortar with nanophase materials, Cem. Concr. Res 34 (2004) 435-438.

[8] H. Li, M. H. Zhang, J. P. Ou, Abrasion resistance of concrete containing nanoparticles for pavement, Wear 260 (2006) 1262-1266.

[9] H. Li, M. H. Zhang, J. P. Ou, Flexural fatigue performance of concrete containing nano-particles for pavement, Int. J. Fatigue 29 (2007) 1292-1301.

[10] L. Senff, J. A. Labrincha, V. M. Ferreira, D. Hotza, W. L. Repette, Effect of nanosilica on rheology and fresh properties of cement pastes and mortars, Constr. Build. Mater. 23 (2009) 2489-2491.

[11] P. Mondal, S. P. Shah, L. Marks, A reliable technique to determine the local mechanical properties at the nanoscale for cementitious materials, Cem. Concr. Res. 37 (2007) 1440-1444.

[12] B. W. Jo, C. H. Kim, G. H. Tae, J. B Park, Characteristics of cement mortar with nano-SiO2 particles. Constr. Build. Mater. 21 (2007) 1351-1355.

[13] R. S. Chen, Q. Ye, Research on the comparison of properties of hardened cement paste between nano-SiO2 and silica fume adde,. Concrete 1 (2002) 7-10.

[14] M. Tang, H. Ba, Y. Li, Study on compound effect of silica fume and nano-SiOx for cement composite materials, J. Chin. Ceram. Soc. 31 (2003) 523-527.

[15] D. F. Lin, M. C. Tsai, The effects of nanomaterials on microstructures of sludge ash cement paste, J. Air Waste Manage. Assoc. 56 (2006) 1146.

[16] J. Y. Shih, T. P. Chang, T. C. Hsiao, Effect of nanosilica on characterization of Portland cement composite, Mater. Sci. Eng. A 424 (2006) 266-274.

[17] T. Ji, Preliminary study on the water permeability and microstructure of concrete incorporating nano-SiO2. Cem. Concr. Res. 35 (2005) 1943-1947.

[18] Jintao Liu, Qinghua Li, Xu. Shilang, Influence of nanoparticles on fluidity and mechanical properties of cement mortar, Constr. Build. Mater. 101 (2015) 892-901.

[19] Yamei Cai, Pengkun Hou, Xin Cheng, Peng Du, Zhengmao Ye, The effect of nanoSiO2 on the properties of fresh and hardened cement based materials through its dispersion with silica fume, Constr. Build. Mater. 148 (2017) 770-780.

[20] R. Moradpour, E. Taheri-Nassaj, T. Parhizkar, M. Ghodsian, The effects of nanoscale expansive agents on the mechanical properties of non-shrink cement-based composites: the influence of nano-MgO addition, Compos. B Eng. 55 (2013) 193-202.

[21] Saloma, A. Nasution, I. Imran, M. Abdullah, Improvement of concrete durability by nanomaterials, Proc. Eng. 125 (2015) 608-612.

[22] E. Mohseni, F. Naseri, R. Amjadi, M.M. Khotbehsara, M.M. Ranjbar, Microstructure and durability properties of cement mortars containing nano-TiO2 and rice husk ash, Constr. Build. Mater. 114 (2016) 656-664.

[23] Morteza Bastami, Mazyar Baghbadrani, Farhad Aslani, Performance of nano- Silica modified high strength concrete at elevated temperatures. Constr. Build. Mater. 68;2014: 656-664.

[24] Jo BW, Kim CH, Tae GH,Park JB. Characteristics of cement mortar with nano silica particles. Constr. Build. Mater. 21 (2007) 1351-1355.

Chapter 5

Simulation and Optimization of an Integrated Process Flow Sheet for Cement Production

Oluwafemi M. Fadayini, Adekunle A. Obisanya, Gloria O. Ajiboye, Clement Madu, Tajudeen O. Ipaye, Taiwo O. Rabiu, Shola J. Ajayi and Joseph T. Akintola

Abstract

In this study the process flow diagram for the cement production was simulated using Aspen HYSYS 8.8 software to achieve high energy optimization and optimum cement flow rate by varying the flow rate of calcium oxide and silica in the clinker feed. Central composite Design (C.C.D) of Response Surface Methodology was used to design the ten experiments for the simulation using Design Expert 10.0.3. Energy efficiency optimization is also carried out using Aspen Energy Analyser. The optimum cement flow rate is found from the contour plot and 3D surface plot to be 47.239 tonnes/day at CaO flow rate of 152.346 tonnes/day and the SiO_2 flow rate of 56.8241 tonnes/day. The R^2 value of 0.9356 determined from the statistical analysis shows a good significance of the model. The overall utilities in terms of energy are found to be optimised by 81.4% from 6.511 x 10^7 kcal/h actual value of 1.211 x 10^7 kcal/h with 297.4 tonnes/day the carbon emission savings.

Keywords: central composite design, optimisation, response surface methodology, cement production, design expert

1. Introduction

Cement is a fine greyish or whitish inorganic, non-metallic powder commonly used as a binding agent in construction materials. It consists of pyroprocessed chemically combined hydraulic cement materials such as calcareous, siliceous, argillaceous and ferriferous [1]. Cement forms paste when mixed with water, which later becomes hard due to cement, mineral hydrate formation when solidified [2]. The various types of cement and their applications such as Portland, Siliceous fly ash, calcareous, slag and Fume silica cement differ by the amount of SiO_2, Al_2O_3, Fe_2O_3, CaO, MgO, SO_3 and other materials such as Na_2O and K_2O composition [3]. Economic growth and urbanisation have made cement one of the most consumed commodity in world with global annual production increase from 3.3 Gt in 2010 to current 4.1 Gt which is still expected to grow moderately in the next decade due to expected infrastructure development in India and other developing Asian and African countries [4, 5]. Cement production consists of three sections: fuel and raw material processing, production of clinker via pyroprocessing and grinding and blending of cement clinker nodules with additive materials such as gypsum and anhydrite for different types of cement types [6]. Natural occurring limestone is ground and mixed in required proportion with silicon and aluminium source such as clay and sand and iron-containing compounds to form a homogenous raw mix called raw meal. The raw meal is then pyroproccessed at a high temperature of about 1450 °C in rotary kiln system where it is dried, preheated, calcined and sintered into cement clinker. The pyroprocessing can be dry, wet, semi-dry or semi-wet and their selection depends on the moisture content of the raw meal, rotary kiln configuration and energy cost. The wet process is cheaper with a high-quality product but very high energy intensity because of the high moisture content of about 36% in raw meal. The dry process is usually more compact with low operational cost and energy consumption compared with the wet process but with lesser product homogeneity [2]. The clinker produced is further grinded with about 5% gypsum which prevents pre-set and controls the hydration rate of the cement. Other types of cement are produced by blending with hydraulic, pozzolanic or inert materials [7]. Cement production processes are energy-intensive and generate huge greenhouse emissions with the clinker energy intensity of about 3.4 GJ/t in 2018 [4] generating about 4% of the global CO_2 emission [8]. Strategies identified to reduce the emissions in cement production include improving heat recovery and energy efficiency [9–11], switching to low carbon source of energy [12], feedstock and material substitute [13–15], reducing the clinker-to-cement ratio [16] and advancing technology innovations such as carbon capture and storage [17, 18]. Cement and concrete technology modelling and simulation have also been used to improve energy efficiency and usage [19–22].

Optimization is a mathematical technique used to find the best solution to objective function (s) by maximising the desired variable and minimising the undesired variables under some set of constraints with the sole aim of improving performance and cost [23]. The optimisation technique in cement and concrete studies can be broadly classified as a meta-heuristic approach and statistical experimental design methods [24]. The meta-heuristic approach is an iterative method that intelligently exploits search space at learning strategies. It includes Genetic Algorithm (GAs), Particle swarm optimization (PSO), Harmony Search (HS), Ant Colony Optimization (ACO), Charged System Search (CSS), Big Bang-Big Crunch (BB-BC), Artificial Bee Colony algorithm (ABC), spherical interpolation of the objective function, Colliding bodies optimization (CBO), Vibrating Particles System (VPS), simulated annealing, krill herd (KH), Whale Optimization Algorithm (WOA), hybrid Harmony Search, force method and genetic algorithm, mine and improved mine blast algorithms [24–26] which are modelled from natural and social behaviours as well as physics laws.

Statistical experimental design methods are widely used to obtain desired optimise solution for a set of constraints [27]. Response Surface Methodology (RSM) is a statistical optimisation technique used to model and analyse a process to determine the effect of independent multivariable on the process response and to evaluate the relations between these variables [28]. RSM is based on understanding the topography of the response surface to determine the most appropriate response region [29]. RSM experimental design can be categorised into Box–Behnken Design (BBD), Central Composite Design (CCD), Dohlert design, Mixture response and three-level factorial design [30–32]. The BBD is created from 3 level factorial design [32] and gives quadratic response model with three minimum number of factors requiring three levels of factors (upper, centre, lower) for each factor and specific positioning of design points [33, 34]. The CCD is developed from the 2 factorial design and gives the quadratic response model with five levels for each factor. Hence it is more robust and insensitive to missing data or experimental runs [34].

In recent years, Response Surface Methodology has been applied to optimise several chemical processes such as extraction [29], adsorption [23], pharmaceutical wastewater treatment [28], leaching [35]. Studies on cement production optimisation have been carried out on clinker simulation using AspenTech [36], cement raw materials blending using a general nonlinear time-varying model [37], cement grinding using population balance model [6], clinker chemistry and kiln energy efficiency using metaheuristic optimization techniques [38], numerical and computational fluid dynamics study of cement calciner [16]. RSM has been efficient and accurate in studies on cement and concrete technology [39–43]. This study focused on the simulation of an integrated wet cement process flow sheet using Aspen HYSYS and optimisation of the cement production rate at minimum raw material feed using CCD of response surface methodology.

2. Methodology

2.1 Cement production simulation

Aspen HYSYS 8.8 was used for the steady-state simulation of the integrated process flow sheet for the cement production [44]. Within the simulation environment, topological optimization (proper arrangement of equipment) was done to enable very high energy savings or optimization. A pure component such as water, CO_2 and air are added as conventional components, while non-conventional components are added as hypothetical components to the HYSYS environment based on their physical properties (molecular weight & density). **Figure 1** shows the block diagram of the production of cement while the HYSYS process flow diagram for the cement production simulation is shown in Figure A1 in the appendix. Limestone is decomposed in the first reactor to give off CO_2 as gas, while the produced CaO is the feed to the Section A reactor to react with silicate to form Calcium disilicate. The produced Calcium disilicate reacts further with the unreacted CaO in reactor B to produce Calcium trisilicate. Calcium oxide (CaO) further reacts with Aluminium oxide to produce tricalcium aluminate, another constituent of cement, while the final product component is produced in section C, where CaO reacts with Aluminate and Ferric oxide to produce tetracalciumaluminoferrite. These separate

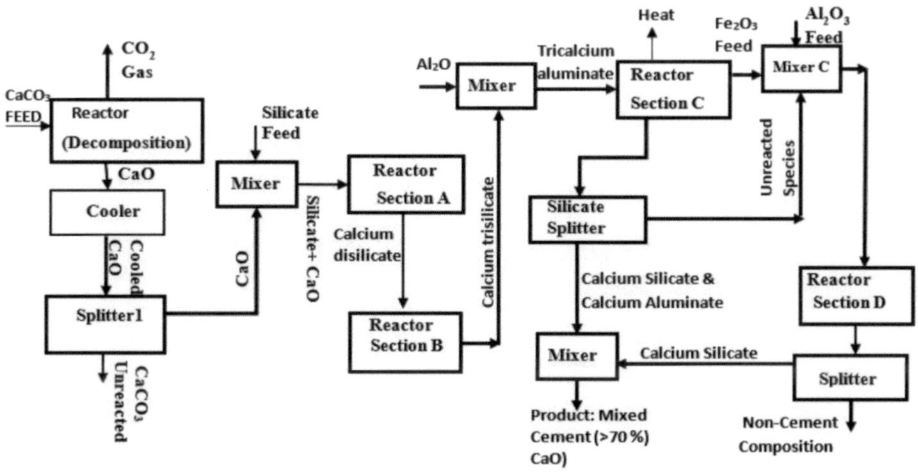

Figure 1.
Process flow diagram for the production of cement.

components produced at a different section of the simulated Kiln are mixed to achieve a matrix compound of the cement product, having over 70% of CaO.

2.2 Aspen Hysys simulation

Aspen Hysys was used for the steady-state simulation of the integrated process flow sheet for the cement production. Within the simulation environment, topological optimization (proper arrangement of equipment) was done to enable very high energy savings or optimization. A pure component such as water, CO_2 and air are added as conventional components, while non-conventional components are added as hypothetical components to the HYSYS environment based on their physical properties (molecular weight and density). Based on the process description, the different reactions taking place in each simulated reactor, as presented in the flowchart are:

$$CaCO_3 \rightarrow CaO + CO_2 (Limestone\ decomposition) \tag{1}$$

$$2CaO + SiO_2 \rightarrow Ca_2SiO_4\ (Section\ A) \tag{2}$$

$$CaO + Ca_2SiO_4 \rightarrow Ca_3SiO_5 (Section\ B) \tag{3}$$

$$3CaO + Al_2O_3 \rightarrow Ca_3Al_2O_6\ (Section\ C) \tag{4}$$

$$4CaO + Al_2O_3 + Fe_2O_3 \rightarrow Ca_4Al_2Fe_2O_{10} (Section\ D) \tag{5}$$

The various products in the various sections of the process reactors are; Tricalcium silicate (Ca_2SiO_4) which is responsible for early strength and the initial set of the cement; Dicalcium silicate (Ca_3SiO_5) which increases the strength as it age; Tricalcium aluminate ($Ca_3Al_2O_6$) which contributes to the concrete strength development in the first few days but least desirable due to its reactiveness with sulphate containing soils and water; Tetracaliumaluminoferrite ($Ca_4Al_2Fe_2O_{10}$) which reduces clinkering temperature. The equipment design parameters employed in this work are provided in **Table 1**.

The flow rate of the major raw materials for the production of cement in the clinkering reactor as depicted by Eqns. (6–12) are carefully chosen based on the standard provided by Winter N. B. [45]. The Chemical parameters based on the oxide composition are very useful in describing clinker characteristics. The following parameters are widely used.

a. *Lime Saturation Factor (LSF):* is the measure of the ratio of alite to belite in the clinker. It is estimated by the ratio of CaO to the sum of other three main oxides SiO_2, Fe_2O_3 and Al_2O_3. The equation is given by:

$$LSF = \frac{CaO}{2.8SiO_2 + 1.2Al_2O_3 + 0.65Fe_2O_3} \tag{6}$$

$$LSF = \frac{190}{2.8(60) + 1.2(15) + 0.65(10)} = 0.98 \tag{7}$$

LSF values in clinkers range between 0.92–0.98. The LSF value of 0.98 falls within an acceptable range.

b. *Silica Ratio (SR):* This is also known as Silica Modulus. The expression of SR is given as:

$$SR = \frac{SiO_2}{Al_2O_3 + Fe_2O_3} \tag{8}$$

Simple separator	Delta P	Stream fractions			
	0.000	Solids in vapour	0.0100		
		Solids in liquid	0.0100		
		Liquid in bottoms	0.0100		
Exchanger	Delta P (bar)	Delta T (°C)	Duty (kcal/h)		
Cooler	0.000	655	2.691 x0 10^7		
Component Splitter	Top Temp. (°C)	Bottom Temp. (°C)	Top Pressure (bar)	Bottom Pressure (bar)	Duty
Splitter 1	30	30	1	1	2.054 x 10^5
Splitter 2	1252	1252	1	1	1.825 x 10^4
Splitter 3	1252	1252	1	1	1.923 x 10^4
Reactors	Delta P	Vessel volume (m^3)	Liquid level (%)	Liquid volume (m^3)	Duty (kcal/h)
$CaCO_3$ decomp reactor	0.0000	0.0000	50.00	25.00	2.6764 x 10^7
Section A reactor	0.0000	50.00	50.00	25.00	7.6209 x 10^6
Section B reactor	0.0000	50.00	50.00	25.00	2.6131 x 10^6
Section C reactor	0.0000	50.00	50.00	25.00	4.5634 x 10^5
Section D reactor	0.0000	50.00	50.00	25.00	7.4721 x 10^5

Table 1.
Equipment design parameter.

Based on the experimental design for the simulated cement production process. The flow rate of Al2O3 and Fe_2O_3 are 15 tonnes/day and 10 tonnes/day respectively. The low level and high levels of SiO_2 are found to be 50 tonnes/day and 60 tonnes/day respectively. Hence, the SR values are the high and low value of the SiO2 flow rate are calculated as follows:

$$SR = \frac{50}{15 + 10} = 2.0 \tag{9}$$

$$SR = \frac{60}{15 + 10} = 2.4 \tag{10}$$

A high silicate ratio means that more calcium silicates are present in the clinker and less aluminate and ferrite. SR is typical, between 2.0 and 3.0. The SR values of 2.0 and 2.4 fall within an acceptable range of 2.0 and 3.0.

c. *Aluminate Ratio (AR):* This is the ratio of aluminate and ferrite phases in the clinker. AR value ranges between 1–4 in Portland clinkers. The flow rate of Al_2O_3 and Fe_2O_3 used in the process simulation are 15 tonnes/day and 10 tonnes/day respectively. The equation governing the AR of the oxide is given by

$$AR = \frac{Al_2O_3}{Fe_2O_3} \tag{11}$$

$$AR = \frac{15}{10} = 1.5 \qquad (12)$$

The mass flow and corresponding clinker quality parameters are presented in **Table 2**.

2.3 Multivariate design of experiment

The central composite design of response surface methodology was used to analyse the effect of CaO and SiO_2 on cement production rate. The total number of experimental runs (N) required for n independent variables and n_c number of replica centre points is given by Eq. 13

$$N = 2^n + 2n + n_c \qquad (13)$$

Design Expert 10.0.3 software was used to generate the experimental design from the ten experimental runs to study the combined effect of two variables on the response. For two variables factor in the experiments; four factorial points (2^n), four axial points ($2n$) and two replicates at the central points (n_c) at distance α = 1.414 from the centre were used for the CCD design. A polynomial empirical model was developed from the ten experimental runs to correlate the response with the independent variables. The mathematical expression can be expressed as:

$$Y = \beta_o + \sum_{i=1}^{n} \beta_i X_i + \sum_{i=1}^{n} \beta_{ii} X_i^2 + \sum_{1 \le i \le j}^{n} \beta_{ij} X_i X_j + \varepsilon \qquad (14)$$

Where Y = Predicted response, β_o = constant coefficient, β_i = Linear coefficient, β_{ii} = Quadratic, β_{ij} = Interaction coefficients and ε = model random error, n the number of variable factors, X_i and X_j are the coded values of the variable parameters [35].

The response generated function distance from the centre $\alpha = 2^{n/4}$. The codes are calculated as a function of the range the factors as shown in **Table 3**.

The central composite experimental design for the synthesis of cement via simulation is depicted in **Table 4**. The mass flow rate of CaO and SiO_2 measured in tonnes/day are the independent variables or predictors which are studied for their effect on the response variable (cement flow rate) at a constant Al_2O_3 and Fe_2O_3 flow rates.

2.4 Model fitting and statistical analysis

The interaction between the variables and the response data as well as the statistical parameters were analysed graphically by analysis of variance (ANOVA) in the Design-Expert software. Regression analysis, significance, F-test, surface and contour plots of the response were also generated from the software. A probability

	Mass flow (tonnes/day)				Clinker quality parameter		
	CaO	SiO_2	Al_2O_3	Fe_2O_3	LSF	SR	AR
High level	190	60	15	10	0.98	2.4	1.5
Low level	135	50	15	10	0.82	2.0	1.5

Table 2.
Raw material mass flow and clinker quality parameter.

	Code	Mathematical relationship
Lower axial point	- α	X_{min}
Lower level	- 1	$[(X_{max} + X_{min})/2] - [\alpha(X_{max} - X_{min})/2]$
Centre point	0	$(X_{max} + X_{min})/2$
Upper level	1	$[(X_{max} + X_{min})/2] + [\alpha(X_{max} - X_{min})/2]$
Upper axial point	α	X_{max}

Table 3.
Relationship between the variable values and their assigned codes.

Factor	Name	Units	Low	High	-α	+α
A	Flow rate of CaO	tonnes/day	135	190	123.609	201.391
B	Flow rate of SiO_2	tonnes/day	50	60	47.9289	62.0711

Table 4.
Design of Experiment using central composite (C.C.D) Design of Response Surface Methodology (R.S.M).

value of 95% confidence level was used to evaluate the significance of the model terms and coefficients.

3. Results and discussion

3.1 Simulation and optimisation of cement flow

All simulations were done in duplicate and the experimental design were generated by the Central Composite Design (C.C.D) of the Design-Expert Software, which resulted in a total of 10 experimental (simulation) runs and the results of the experiments (simulations) are shown in **Table 5**.

Run No.	Factors				Response	
	Coded levels		Actual values		The flow rate of cement	
	A	B	A	B	HYSYS Simulation	C.C.D Model
1	1.000	1.000	190	60	46.6	45.6674
2	−1.000	−1.000	135	50	30.74	31.9851
3	0.000	0.000	162.5	55	46.04	46.04
4	0.000	−1.414	162.5	47.9289	30.57	31.271
5	1.000	−1.000	190	50	44.3	42.1551
6	−1.414	0.000	123.609	55	46.04	43.4867
7	0.000	0.000	162.5	55	46.04	46.04
8	−1.000	1.000	135	60	46.6	49.0574
9	1.414	0.000	201.391	55	46.04	48.2808
10	0.000	1.414	162.5	62.0711	46.84	45.8265

Table 5.
Simulation and predicted results from central composite design (C.C.D).

The change in mean response in cement flow per unit increase in variable occurs when other predictors area kept constant and is estimated by the coefficient of estimation and is presented **Table 6**.

The empirical quadratic equation for the optimal cement product rate as a function of CaO and SiO_2 mass flow in coded form as derived from **Table 7** was obtained according to the CCD and is given in Eq. 15

$$C = 46.04 + 1.70A + 5.15B - 3.39AB - 0.078A^2 - 3.75B^2 \quad (15)$$

The test of the significance and the adequacy of the model and its coefficients lack fitness which was based on F-value or P-value at 95% confidence level was tested from analysis of variance (ANOVA) and the result is presented in **Table 7**. The result shows that the model at an F value of 20.35 and a very low P-value of 0.0005 indicates that the statistical regression model was significant. The result also shows that A, B, AB and B^2 are significant terms.

The regression statistical analysis is summarised in **Table 8**. The R squared value of 0.9356 is in good agreement with the adjusted R-square value of 0.8896, showing a good fit of the model, as the closer the R squared value to 1.00, the more significant the model. The adequate precision of 13.55 indicates the low noise level and a

	Coefficient		**Standard**	**95% CI**	**95% CI**	
Factor	Estimate	df	Error	Low	High	VIF
Intercept	46.04	1	0.86	44.00	48.08	
A-Flowrate of CaO	1.70	1	0.68	0.080	3.31	1.00
B-Flow rate of Silica	5.15	1	0.68	3.53	6.76	1.00
AB	−3.39	1	0.97	−5.67	−1.11	1.00
A^2	−0.078	1	0.73	−1.81	1.65	1.02
B^2	−3.75	1	0.73	−5.48	−2.01	1.02

Table 6.
Coefficient estimation for cement flow rate in terms of coded factors.

Source	**Sum of squares**	**df**	**Mean**	**F value**	**p-value (Prob > F)**
Model	379.61	5	75.92	20.35	0.0005
A	*22.98*	*1*	*22.98*	*6.16*	*0.0421*
B	*211.86*	*1*	*211.86*	*56.78*	*0.0001*
AB	*45.97*	*1*	*45.97*	*12.32*	*0.0099*
A^2	*0.042*	*1*	*0.042*	*0.011*	*0.9180*
B^2	*97.60*	*1*	*97.60*	*26.16*	*0.0014*
Residual	26.12	7	3.73		
Lack of Fit	26.12	*3*	*8.71*		
Pure Error	0.000	*4*	0.000		
Cor. Total	405.73	12			

Table 7.
ANOVA results for the statistical model for cement flow rate.

Parameter	Values
R-Squared	0.9356
Adj R-Squared	0.8896
Pred R-Squared	0.5422
Mean	43.69
Std. Dev.	1.93
C.V. %	4.42
Adeq Precision	13.554

Table 8.
Statistical information for the statistical model for cement flow rate.

strong signal for optimisation. Hence, this indicates the two predictors (flow rate of CaO and flow rate of SiO_2) could predict the flow rate of cement, thus the model equation, contour plot and 3D surface plot could be used to predict the response (flow rate of the cement).

3.2 Surface plots

The contour plot which shows the possible relationship between the CaO, SiO_2 and cement product mass flow is presented in **Figure 2**. The darker red regions indicate higher C (response) values. Here, the optimum flow rate of cement is found from the isolines to be 47.748 tonnes/day at a flow rate of 152.346 tonnes/day of CaO and 56.8241 tonnes/day SiO_2.

The three-dimensional (3D) response surface plots obtained from the model equation using Design Expert 10.03 is depicted in **Figure 3**. This depicts the effect

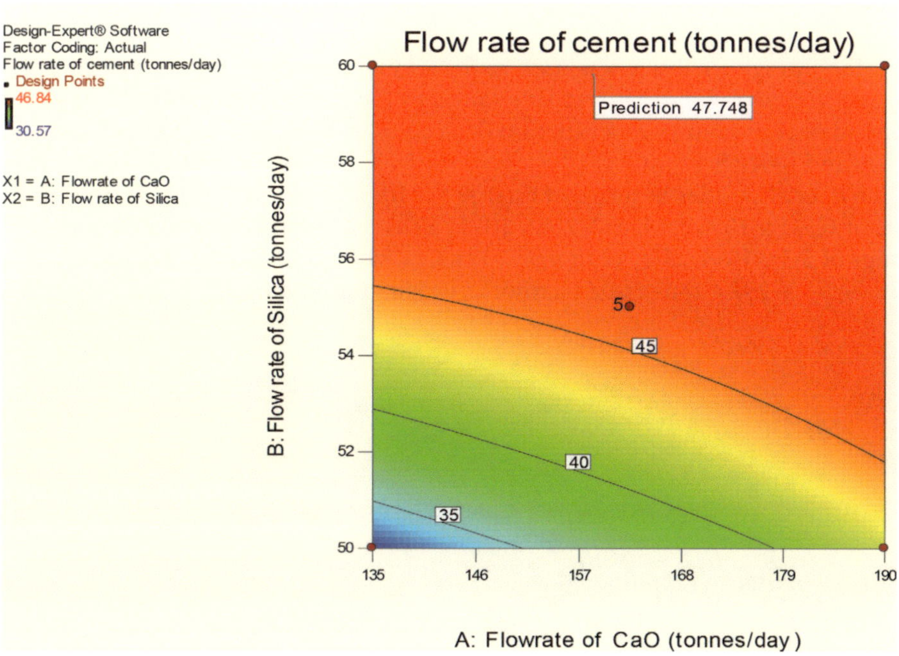

Figure 2.
Contour surface plot showing the effects of the flow rate of CaO and flow rate of SiO_2 on the flow rate of cement.

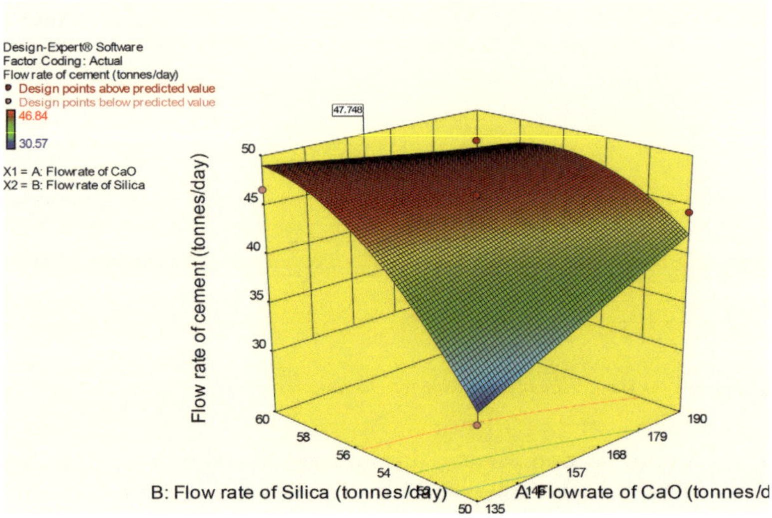

Figure 3.
Response surface plot of the effects of CaO and SiO_2 mass flow on cement production.

of the flow rate of CaO, the flow rate of SiO_2 on the flow rate of cement. The flow rate of cement was observed to increase with an increase in the flow rate of CaO. Conversely, increasing the flow rate of SiO_2 did not increase the flow rate of cement. Hence, the major predictor in cement production in the clinkering section is the flow rate of CaO.

The optimization plot of the cement output is shown in **Figure 4**. The optimum cement flow rate of 47.239 tonnes/day is found to be at CaO flow rate of 152.346 tonnes/day and SiO_2 flow rate of 56.8241 tonnes/day.

3.3 Energy optimization

Aspen Energy Analyser was used to determine the percentage of energy savings based on converged steady-state simulation of the process flow sheet in **Figure 1**. The total energy savings as a function of process utilities and carbon emissions are present in **Figure 5**.

The overall utilities in terms of energy are found to be optimised from the actual value of 6.511×10^7 kcal/h to 1.211×10^7 kcal/h and indicating available energy savings of 5.3×10^7 kcal/h, with overall energy savings of 81.40% which also correspond to 297.4 tonnes/day carbon emission reduction.

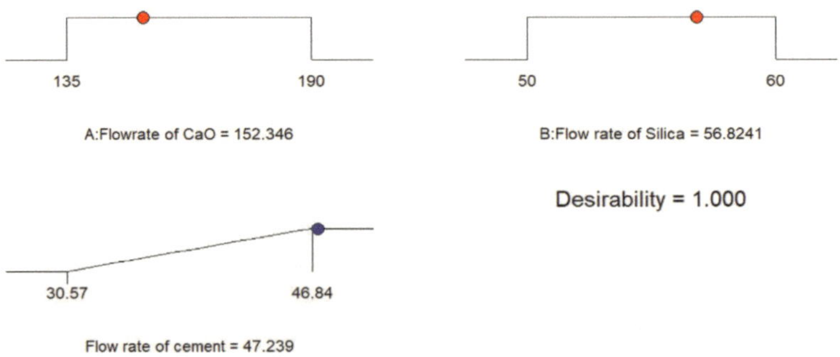

Figure 4.
Optimization plot showing the effects of the flow rate of CaO and flow rate of SiO_2 on the flow rate of cement.

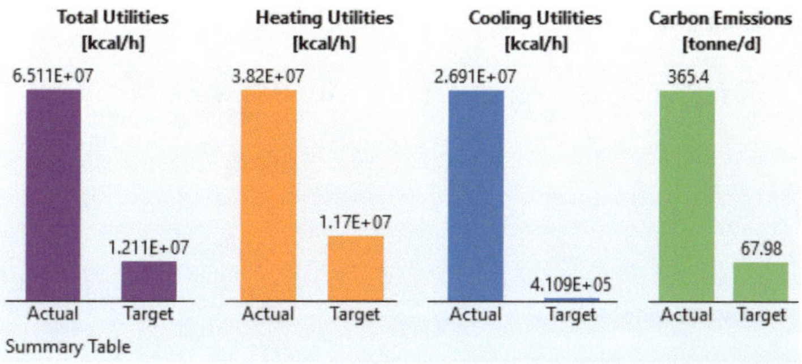

Figure 5.
Energy savings at optimum feed (CaO and SiO_2) rate and product (cement) flow rate.

4. Conclusion

Process flow diagram for the cement production was simulated to achieve high energy optimization and optimum cement flow rate by minimising the flow rate of the feed (CaO and SiO_2). Central composite Design (C.C.D) of Response Surface Methodology used to design the experiment for the simulation using Design Expert 10.0.3. The optimum cement flow rate is found from surface and contour plots to be 47.239 tonnes/day at CaO flow rate of 152.346 tonnes/day and SiO_2 flow rate of 56.8241 tonnes/day. The R squared value of 0.9356 determined from the statistical analysis shows a very high significance of the model. Energy efficiency optimization is also carried out using Aspen Energy Analyser. The overall utilities in terms of energy are found to be optimised by 81.4 % from 6.511×10^7 kcal/hactual value to 1.211×10^7 kcal/h with 297.4 tonnes/day the carbon emission savings.

Further work could be performed on fault identification and diagnosis of the process plant. Incorporated with an automated plant to guarantee the safety of workers, reduce environmental problems and increase yield to sustain production improvement.

5. Recommendations

This research work sought to recommend the following concerns in which cement production could be improved:

1. Research and development (R&D) in the cement production, processing and utilisation should be encouraged. This will play a vital role in the construction industry, operation and maintenance of efficient road network and effective transportation system.

2. Automation of cement and kiln sections of the cement production is recommended

3. Optimization of the cement production can be tailored into the fabrication of high tech cement equipment and parts.

4. Optimization of the limestone crusher to quantify the amount of crushed limestone is needed.

5. Looking into future the results obtained in this research will open up several important possibilities in the cement production at optimum conditions. This will have a multiplier effect on infrastructural amenities development.

Conflict of interest

There is no conflict of interest associated with this work.

Appendix

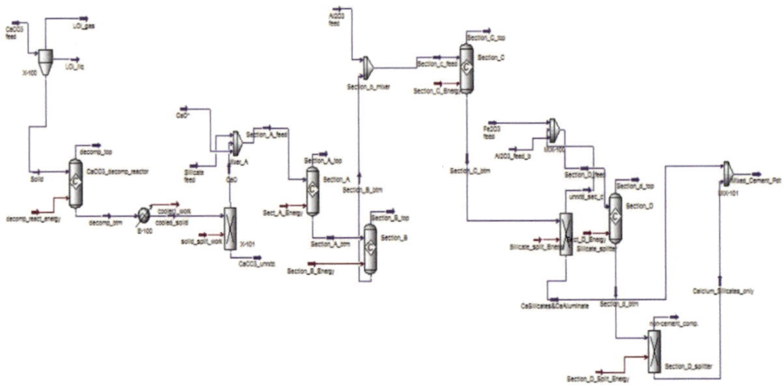

Figure A1.
HYSYS process flow simulation diagram for the production of cement.

Author details

Oluwafemi M. Fadayini[1*], Adekunle A. Obisanya[2], Gloria O. Ajiboye[2], Clement Madu[1], Tajudeen O. Ipaye[3], Taiwo O. Rabiu[4], Shola J. Ajayi[1] and Joseph T. Akintola[1]

1 Department of Chemical Engineering, Lagos State Polytechnic, Ikorodu, Lagos, Nigeria

2 Department of Chemical Engineering, Yaba College of Technology, Yaba, Lagos, Nigeria

3 Department of Civil Engineering, Lagos State Polytechnic, Ikorodu, Lagos, Nigeria

4 Department of Mechanical Engineering, Lagos State Polytechnic, Ikorodu, Lagos, Nigeria

*Address all correspondence to: olufeday@gmail.com; fadayini.o@mylaspotech.edu.ng

References

[1] EPA, "Mineral Products Industry," in *Compilation of air pollutant emission factors. Volume I: stationary point and area sources*, AP-42 5th., Research Triangle Park, NC: U.S. Environmental protection agency, 1995.

[2] N. Martin, E. Worrell, and L. Price, "Energy Efficiency and Carbon Dioxide Emissions Reduction Opportunities in the U.S. Cement Industry," Berkeley, 1999. doi: 10.2172/751775.

[3] S. P. Dunuweera and R. M. G. Rajapakse, "Cement Types, Composition, Uses and Advantages of Nanocement, Environmental Impact on Cement Production, and Possible Solutions," *Adv. Mater. Sci. Eng.*, vol. 2018, 2018, doi: 10.1155/2018/4158682.

[4] IEA, "Cement," Paris, 2020. [Online]. Available: https://www.iea.org/reports/cement.

[5] TERI, "CEMENT INDUSTRY; Trends Report," New Delhi, 2017. [Online]. Available: http://www.teriin.org/library/files/Cement-Industry-Trends-Report2017.pdf.

[6] A. Jankovic, W. Valery, and E. Davis, "Cement grinding optimisation," *Miner. Eng.*, vol. 17, no. 11–12, pp. 1075–1081, 2004, doi: 10.1016/j.mineng.2004.06.031.

[7] D. Olsen, S. Goli, D. Faulkner, and A. Mckane, "Opportunities for Energy Efficiency and Demand Response in the California Cement Industry," no. December. Lawrence Berkeley National Laboratory, 2010.

[8] J. G. J. Olivier and J. A. H. W. Peters, "Trends in Global CO2 and Total Greenhouse Gas Emissions: Report 2019," The Hague, 2020. [Online]. Available: www.pbl.nl/en.

[9] H. Mikulčić, M. Vujanović, N. Markovska, R. V. Filkoski, M. Ban, and N. Duić, "CO2 emission reduction in the cement industry," *Chem. Eng. Trans.*, vol. 35, pp. 703–708, 2013, doi: 10.3303/CET1335117.

[10] S. Zhang, H. Ren, W. Zhou, Y. Yu, and C. Chen, "Assessing air pollution abatement co-benefits of energy efficiency improvement in cement industry: A city-level analysis," *J. Clean. Prod.*, vol. 185, pp. 761–771, 2018, doi: 10.1016/j.jclepro.2018.02.293.

[11] M. J. S. Zuberi and M. K. Patel, "Bottom-up analysis of energy efficiency improvement and CO2 emission reduction potentials in the Swiss cement industry," *J. Clean. Prod.*, vol. 142, pp. 4294–4309, 2017, doi: 10.1016/j.jclepro.2016.11.178.

[12] N. Chatziaras, C. S. Psomopoulos, and N. J. Themelis, "Use of waste-derived fuels in cement industry: a review," *Manag. Environ. Qual. An Int. J.*, vol. 27, no. 2, pp. 178–193, 2016, doi: 10.1108/MEQ-01-2015-0012.

[13] E. Marchetti, "Use of Agricultural Wastes as Supplementary Cementitious Materials," KTH ROYAL INSTITUTE OF TECHNOLOGY, 2020.

[14] A. Naqi and J. G. Jang, "Recent progress in green cement technology utilizing low-carbon emission fuels and raw materials: A review," *Sustain.*, vol. 11, no. 2, 2019, doi: 10.3390/su11020537.

[15] R. Maddalena, J. J. Roberts, and A. Hamilton, "Can Portland cement be replaced by low-carbon alternative materials? A study on the thermal properties and carbon emissions of innovative cement," *J. Clean. Prod.*, vol. 186, no. April, pp. 933–942, 2018, doi: 10.1016/j.jclepro.2018.02.138.

[16] H. Mikulcic, M. Vujanovic, and N. Duic, "Improving the Sustainability of Cement Production by Using Numerical

Simulation of Limestone Thermal Degradation and Pulverized Coal Combustion in a Cement Calciner," *J. Clean. Prod.*, vol. 88, pp. 262–271, 2015.

[17] P. Markewitz *et al.*, "Carbon capture for CO2 emission reduction in the cement industry in Germany," *Energies*, vol. 12, no. 12, pp. 1–27, 2019, doi: 10.3390/en12122432.

[18] WSP Parson Brinkerhoff and DNV GL, "Industrial Decarbonisation & Energy Efficiency Roadmaps to 2050: Cement," 2015.

[19] C. Csernyei and A. G. Straatman, "Numerical modelling of a rotary cement kiln with improvements to shell cooling," *Int. J. Heat Mass Transf.*, vol. 102, pp. 610–621, 2016, doi: 10.1016/j.ijheatmasstransfer.2016.06.058.

[20] H. R. Goshayeshi and F. K. Poor, "Modeling of Rotary Cement Kiln (In Persian)," *Energy Power Eng.*, vol. 8, pp. 23–33, 2016.

[21] S. Sadighi, M. Shirvani, and A. Ahmad, "Rotary cement kiln coating estimator: Integrated modelling of the kiln with shell temperature measurement," *Can. J. Chem. Eng.*, vol. 89, no. 1, pp. 116–125, Feb. 2011, doi: 10.1002/cjce.20365.

[22] K. V. Sabarish, M. Akish Remo, and P. Paul, "Optimizing the Concrete Materials by Taguchi Optimization Method," *IOP Conf. Ser. Mater. Sci. Eng.*, vol. 574, no. 1, 2019, doi: 10.1088/1757-899X/574/1/012002.

[23] S. Saini, J. Chawla, R. Kumar, and I. Kaur, "Response surface methodology (RSM) for optimization of cadmium ions adsorption using - C 16–6 - 16 incorporated mesoporous MCM - 41," *SN Appl. Sci.*, vol. 1, no. 8, pp. 1–10, 2019, doi: 10.1007/s42452-019-0922-5.

[24] C. TAO, "Optimization of Cement Production and Hydration for Improved," UNIVERSITY OF FLORIDA, 2017.

[25] F. Omidinasab and V. Goodarzimehr, "A hybrid particle swarm optimization and genetic algorithm for truss structures with discrete variables," *J. Appl. Comput. Mech.*, vol. 6, no. 3, pp. 593–604, 2020, doi: 10.22055/JACM.2019.28992.1531.

[26] L. J. Li, Z. B. Huang, and F. Liu, "A heuristic particle swarm optimization method for truss structures with discrete variables," *Comput. Struct.*, vol. 87, no. 7–8, pp. 435–443, Apr. 2009, doi: 10.1016/j.compstruc.2009.01.004.

[27] M. J. Simon, "Concrete Mixture Optimization Using Statistical Methods: Final Report," 2003. [Online]. Available: http://www.fhwa.dot.gov/publications/research/infrastructure/pavements/03060/03060.pdf.

[28] S. Barisci and O. Turkay, "Optimization and modelling using the response surface methodology (RSM) for ciprofloxacin removal by electrocoagulation," *Water Sci. Technol.*, vol. 73, no. 7, pp. 1673–1679, 2016, doi: 10.2166/wst.2015.649.

[29] A. Y. Aydar, "Utilization of Response Surface Methodology in Optimization of Extraction of Plant Materials," in *Statistical Approaches With Emphasis on Design of Experiments Applied to Chemical Processes*, V. Silva, Ed. InTech, 2018, pp. 157–168.

[30] S. A. Adeyeye, "Banana Drying Kinetics," in *Banana Nutrition - Function and Processing Kinetics*, IntechOpen, 2020, pp. 0–20.

[31] M. N. Chollom, S. Rathilal, F. M. Swalaha, B. F. Bakare, and E. K. Tetteh, "Comparison of response surface methods for the optimization of an up-flow anaerobic sludge blanket for the treatment of slaughterhouse wastewater," *Environ. Eng. Res.*, vol. 25,

no. 1, pp. 114–122, 2020, doi: 10.4491/eer.2018.366.

[32] A. I. Khuri, "Response Surface Methodology and Its Applications In Agricultural and Food Sciences," *Biometrics Biostat. Int. J.*, vol. 5, no. 5, pp. 155–163, 2017, doi: 10.15406/bbij.2017.05.00141.

[33] B. Olawoye, *A comprehensive handout on central composite design (CCD)*, no. July. 2016.

[34] Shari, "Choosing the Best Design for Process Optimization," 2017. https://www.statease.com/blog/choosing-best-design-process-optimization/ (accessed Nov. 27, 2020).

[35] S. Kumar, H. Meena, S. Chakraborty, and B. C. Meikap, "International Journal of Mining Science and Technology Application of response surface methodology (RSM) for optimization of leaching parameters for ash reduction from low-grade coal," *Int. J. Min. Sci. Technol.*, vol. 28, no. 4, pp. 621–629, 2018, doi: 10.1016/j.ijmst.2018.04.014.

[36] B. Hokfors, M. Eriksson, and E. Viggh, "Modelling the cement process and cement clinker quality," *Adv. Cem. Res.*, vol. 26, no. 6, pp. 311–318, 2014, doi: 10.1680/adcr.13.00050.

[37] X. Li, H. Yu, and M. Yuan, "Modeling and Optimization of Cement Raw Materials Blending Process," *Math. Probl. Eng.*, vol. 2012, 2012, doi: 10.1155/2012/392197.

[38] C. Tao, "Optimization of cement production and hydration for improved performance, energy conservation, and cost," UNIVERSITY OF FLORIDA, 2017.

[39] G. Cibilakshmi and J. Jegan, " A DOE approach to optimize the strength properties of concrete incorporated with different ratios of PVA fibre and nano-Fe 2 O 3," *Adv. Compos. Lett.*, vol. 29, p. 2633366X2091388, 2020, doi: 10.1177/2633366x20913882.

[40] T. F. Awolusi, O. L. Oke, O. O. Akinkurolere, and O. D. Atoyebi, "Comparison of response surface methodology and hybrid-training approach of artificial neural network in modelling the properties of concrete containing steel fibre extracted from waste tyres," *Cogent Eng.*, vol. 6, no. 1, 2019, doi: 10.1080/23311916.2019.1649852.

[41] M. Sonebi and M. T. Bassuoni, "Investigating the effect of mixture design parameters on pervious concrete by statistical modelling," *Constr. Build. Mater.*, vol. 38, pp. 147–154, Jan. 2013, doi: 10.1016/j.conbuildmat.2012.07.044.

[42] K. E. Alyamac, E. Ghafari, and R. Ince, "Development of eco-efficient self-compacting concrete with waste marble powder using the response surface method," *J. Clean. Prod.*, vol. 144, pp. 192–202, Feb. 2017, doi: 10.1016/j.jclepro.2016.12.156.

[43] L. Soto-Pérez, V. López, and S. S. Hwang, "Response Surface Methodology to optimize the cement paste mix design: Time-dependent contribution of fly ash and nano-iron oxide as admixtures," *Mater. Des.*, vol. 86, pp. 22–29, Dec. 2015, doi: 10.1016/j.matdes.2015.07.049.

[44] R. C. Carpio, L. D. S. Coelho, R. J. Silva, and A. B. Jorge, "Case Study in Cement Kilns Alternative Secondary Fuels Mixing Using Sequential Quadratic Programming, Genetic Algorithms, and Differential Evolution," 2005, doi: 10.1590/S1678-58782008000400010.

[45] N. Winter, "Understanding Cement," 2005. https://www.understanding-cement.com/clinker.html (accessed Nov. 02, 2020).

Chapter 6

Peculiarities of Portland Cement Clinker Synthesis in the Presence of a Significant Amount of SO_3 in a Raw Mix

Oleg Sheshukov and Michael Mikheenkov

Abstract

Due to the depletion of the raw material base and a technogenic materials addition into a raw mix for the Portland cement clinker synthesis, sulfur and its oxides amount in a raw mix increases. According to literature the Portland cement clinker synthesis in the presence of a sulfur oxides significant amount is difficult. As the content of SO_3 in the raw mix increases the amount of C_2S increases while C_3S and C_3A amount decrease. With an equal total content of C_2S and C_3S in the clinker their ratio C_3S/C_2S decreases with an increased content of SO_3. These factors lead to a deterioration in the Portland cement clinker quality. The clinker formation reactions thermodynamic analysis and some experimental studies allow determining reasons for the Portland cement clinker quality deterioration. It was found that the presence significant amount of a SO_3 in the raw mix the synthesis in solid phase of low-basic $C_4A_3\overline{S}$ (ye'elimite) is the thermodynamically preferred rather than high-basic C_3A and C_4AF. As a result, excess and crystallized free lime inhibits the C_3S synthesis through the liquid phase. The experimental studies result helped to develop a methodology for calculating the composition of a raw mix from materials with significant amount of SO_3. It allows to reduce the SO_3 negative effect on the Portland cement clinker synthesis and to obtain high-quality Portland cement.

Keywords: high-sulfate raw material, Portland cement clinker, alite, belite, brownmillerite, ye'elimite, calculation procedure

1. Introduction

Sulfur and its oxides in the form of sulfate and sulfide minerals appear in the raw mixture of Portland cement clinker with the main raw materials for the preparation of clinker, namely clay and carbonate rock as well as additives and fuel. The technogenic origin additives of the metallurgical and heat-power industry, namely slags, fly ashes and oil cokes have especially high concentration of sulfur compounds [1–3]. According to [4] sulfur with calcium oxide forms calcium sulfate $CaSO_4$ under conditions of oxidative burning. Depending on the burning temperature the latter with the alkaline components of the raw mix forms alkaline metal sulfates or with clinker minerals forms sulfospurrit $2(C_2S)C\overline{S}$ and ye'elimite $C_4A_3\overline{S}$ and participates in the alkali-sulfate cycle of the furnace.

When the SO_3 content in the clinker is less than 2.0% it has a positive effect on the synthesis of Portland cement clinker since the alkali metal sulfates formed in its presence are effective melts that reduce the temperature of appearance of the liquid phase and its viscosity providing accelerated synthesis of clinker minerals [5, 6].

The positive role of SO_3 in clinker is also evident when using a raw mixture with a significant content of alkalis. When the molar ratio of $SO_3/(Na_2O + K_2O)$ is close to 1, the excess alkali is removed from the raw material mixture due to the removal of alkali metal sulfates during heating [4, 6].

If the SO_3 content in the clinker exceeds 2.0%, there are negative phenomena associated with both the quality of Portland cement clinker and its production technology.

According to [7, 8] when the content of SO_3 in the raw mixture increases the amount of C_2S increases, C_3S decreases and when the total content of these phases in the clinker is equal, their C_3S/C_2S ratio decreases.

Since alite (C_3S) is the most active and refractory component of Portland cement clinker the alite reduction leads to a decrease of the clinker refractoriness and activity. A decrease of the clinker refractoriness appears in the formation of rings in the furnace and incrustation in the calcinators cyclones and a decrease in activity shows in the drop in the strength of Portland cement.

It was found [9, 10] that not only the C_3S content but also C_3A content decrease in high-sulfate clinker. The reason for the decrease in the C_3A content in high-sulfate clinker is the isomorphic substitution of silicon for aluminum in silicate phases.

The reasons for the decrease of alite (C_3S) content in Portland cement clinker with an increase of SO_3 content in it are not found in the literature.

The main purpose of this study is to determine the reasons for the SO_3 negative impact on the Portland cement clinker synthesis and to develop a method to prevent it.

To achieve this goal, it is needed to:

- analyze literary sources;
- perform thermodynamic calculations;
- conduct experimental research;
- determine the reasons for the SO_3 negative influence on the Portland cement clinker synthesis;
- develop methods to prevent that negative impact.

2. Materials and experiment methods

Raw mixtures were burnt to produce Portland cement clinker with following modular characteristics: LSF = 0.92, n = 2.3 and p = 1.69. For the preparation of Portland cement clinker a raw mix based on limestone, clay, quartz sand and natural gypsum was used. When preparing the raw mix the composition was modeled by the introduction of gypsum into the raw mix with raw components in the ratio: raw mix/gypsum = 95/5%, to achieve SO_3 in the raw mix of more than 2.0%. The raw mix was homogenized in a laboratory mill by joint grinding of raw components and gypsum for 30 minutes. The homogeneous mixture was moistened and pressed at a pressure of 50 MPa.

Material	CaO	SiO_2	Al_2O_3	Fe_2O_3	SO_3	MgO	LOI	Total
Limestone ($CaCO_3$)	56.35	0.04	0.06	0.04	0.03	0.37	42.8	99.69
Clay	1.21	50.70	19.90	11.80	0	1.40	14.81	99.90
Quartz sand	0.03	98.80	0.45	0.03	0	0	0.13	99.40
Natural gypsum	32.02	0.80	0.45	0.23	44.50	0	22.0	100.0
Clinker without gypsum	67.17	22.19	6.08	3.59	0.035	0.85	0	99.99
Clinker with gypsum	65.25	21.12	5.80	3.42	2.26	0.80	1.1	99.99

Table 1.
The chemical composition of raw materials and clinker, mass %.

The samples were burnt at temperatures from 1100 to 1300 °C. The synthesized clinker was crushed to a residue on the sieve No. 008 no more than 5.0% and the sugar-glycerate method was used to determine the content of CaO_{free} in it. The phase composition of fully synthesized clinkers was determined by chemical and x-ray methods. Qualitative x-ray phase analysis was performed using an XRD-7000 diffractometer (Shimadzu). Quantitative x-ray phase analysis was performed using a STADI-P diffractometer (STOE, Germany). The shooting was made under CuKα-radiation (40 kV, 30 mA), with graphite monochromator, in the range of scattering angles 2Θ = 10–70 deg., with a step of 0.02 deg. and an excerpt of 2 s. The results were analyzed using the PDF-2 database (Release 2008 RDB 2.0804). Thermal analysis (TA) of hydration products was performed using the DSC method (differential scanning calorimetry) on the STA 449 F3 Jupiter thermal analyzer (Netzsch-Geratebau GmbH). The temperature varied from room temperature (approximately 20°C) to 800°C at a heating rate of 10°C/min. The samples of the hydration products prepared by a grinding of synthesized clinker and on their basis cement paste prepared. The size of cement pastes were 20 x 20 x 20 mm, which were prepared under the condition of water: cement ratio of 0.4. After preparation, all samples were placed in box with water at temperature of 20°C. In a box samples were maintained 28 day before full hydration. Chemical analysis of cement was performed in accordance with the requirements of Russian Standard 5382–91. For reception of micro-photos the optical microscope Olympus GX-51 (Japan) was used. Samples of clinker was polished and their surface was etching by Nital [11]. After this procedure alite was painted in green-violet color, and belite in light brown color.

The chemical composition of raw materials and clinker is shown in **Table 1**.

3. Experimental results and discussion

The results of qualitative phase analysis of clinker based on a raw mix with the addition of 5% gypsum are shown in **Figure 1**.

Figure 2 shows a micrograph of clinker.

The content of free lime in a clinker based on a raw mix with the addition of 5% gypsum is shown in **Table 2**.

Analysis of the data presented in **Figures 1** and **2** and **Table 2** shows that the introduction of 5% gypsum into the raw mix suppresses the alite formation and contributes to the formation of a significant amount of free lime in the clinker. The diffraction peaks which are typical for free lime are present in a clinker based on a raw mix with the addition of 5% gypsum burnt at a temperature of 1300°C and the diffraction peaks which are typical for alite are not fixed at this temperature.

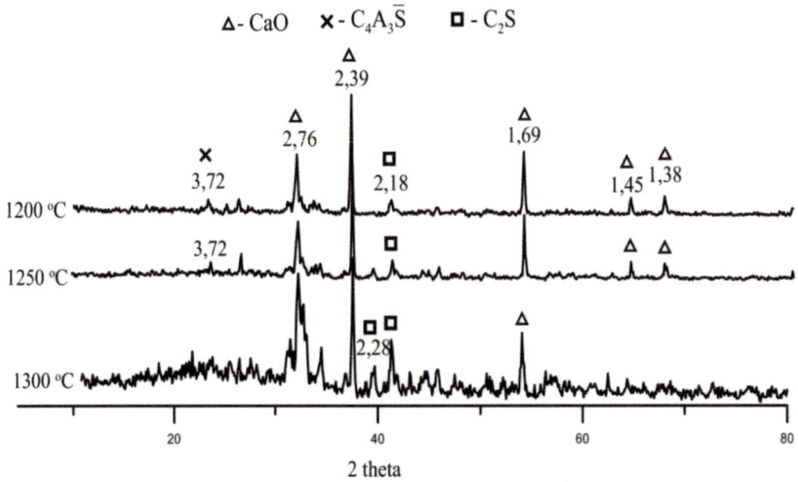

Figure 1.
The results of qualitative phase analysis of clinker based on a raw mix with the addition of 5% gypsum.

Figure 2.
A micrograph of clinker based on a raw mix with the addition of 5% gypsum.

Material	The content of free lime in a clinker, mass %, at the burning temperature, °C		
	1100	1200	1300
Clinker based on a raw mix with the addition of 5% gypsum	14.0	13.84	11.6

Table 2.
The content of free lime in a clinker.

Clinker completely melts when it is heated to a temperature of 1350°C. Analysis of the melt heated to a temperature of 1600°C indicates that alite is formed during this overheating but there is 6.2% free lime in the clinker.

The reason for suppressing the alite formation in a high-sulfate clinker is likely the formation of a well-crystallized CaO_{free} which is not soluble in the liquid phase, does not interact with C_2S and does not form C_3S. The absence of alite in the clinker reduces its refractoriness and contributes to the appearance of a melt at temperatures below the temperatures of Portland cement clinker synthesis.

The formation of a significant amount of free lime in the clinker during the decomposition of gypsum is unlikely because with the stoichiometric ratio of CaO and SO_3 in gypsum anhydrite, respectively 41.19% and 58.81%, 2.06% of CaO_{free} can be formed from the decomposition of 5.0% of gypsum and according to the data presented in **Table 2**, more than 11% of CaO_{free} is formed at a temperature of 1300°C.

To determine the reasons for the appearance of a significant amount of free lime in a high-sulfate clinker a thermodynamic analysis of the reactions of C_3A and C_4AF formation was performed according to the data of [12] as well the reactions of $C_4A_3\overline{S}$ formation presented in [13]. The analysis results are presented in **Table 3**.

According to the data presented in **Table 4**, the synthesis of C_3A and C_4AF in a low-sulfate clinker is thermodynamically possible from the simple minerals, namely CaO, Al_2O_3 and Fe_2O_3 (reactions No. 1–2). In a high-sulfate clinker, reaction of the formation of ye'elimite $C_4A_3\overline{S}$ (reaction No. 3) is thermodynamically preferable since the free Gibbs energy of it is more negative than that of reactions No. 1 and 2.

There are different opinions about the formation mechanism of ye'elimite $C_4A_3\overline{S}$ in clinker during heating. According to [13] the synthesis of ye'elimite due to an excess of lime at the time of its formation begins with the formation of mayenite according to the scheme $C_{12}A_7 \rightarrow CA \rightarrow C_3A_3C\overline{S}$. According to our data [14] in the pressed raw mix due to its higher reaction ability the synthesis of ye'elimite proceeds according to the scheme $CA_2 \rightarrow CA \rightarrow C_2A_2 \rightarrow C_3A_3 \rightarrow C_3A_3C\overline{S}$ with the formation of calcium monoaluminate (CA) at temperatures about 700°C and its presence throughout the burning temperature range up to 1300°C. In the presence of calcium monoaluminate the formation of C_3A and C_4AF is thermodynamically impossible (reactions No. 4–6). Based on these studies the authors of [12] concluded that in the presence of calcium monoaluminate C_3A and C_4AF are formed not in solid-phase synthesis but from a melt.

Therefore if the calculation of the raw mix is carried out according to the usual scheme for the formation of C_3A and C_4AF minerals in a high-sulfate clinker, and in fact in a high-sulfate clinker low-base aluminates and calcium ferrites are formed before the melt appears, then due to the difference in the lime content in these minerals, free lime can be formed in the clinker by reactions:

$$3CaO + Al_2O_3 \rightarrow CaO \cdot Al_2O_3 + 2CaO, \quad (1)$$

$$4CaO + Al_2O_3 + Fe_2O_3 \rightarrow CaO \cdot Al_2O_3 + CaO \cdot Fe_2O_3 + 2CaO. \quad (2)$$

No.	Reactions	The value of ΔG_o, kJ/mol at temperature, K				
		298	1023	1200	1400	1500
1	$3CaO + Al_2O_3 = 3CaO{\cdot}Al_2O_3$	−17.0	−41.8	−47.0	−52.9	−55.7
2	$4CaO + Al_2O_3 + Fe_2O_3 = 4CaO{\cdot}Al_2O_3{\cdot}Fe_2O_3$	−49.3	−64.9	−64.1	−60.83	−58.1
3	$3CaO + 3Al_2O_3 + CaSO_4 = 3CaO{\cdot}3Al_2O_3{\cdot}CaSO_4$	−99.1	−445.1	−583.6	−758.9	−853.5
4	$CaO{\cdot}Al_2O_3 + 2CaO = 3CaO{\cdot}Al_2O_3$	+33.7	+32.3	+31.7	+33.26	+34.4
5	$CaO{\cdot}Al_2O_3 + CaO + 2CaO{\cdot}Fe_2O_3 = 4CaO{\cdot}Al_2O_3{\cdot}Fe_2O_3$	+10.4	+39.7	+49.0	+60.3	+66.2
6	$CaO{\cdot}Al_2O_3 + 2CaO + CaO{\cdot}Fe_2O_3 = 4CaO{\cdot}Al_2O_3{\cdot}Fe_2O_3$	+42.4	+72.1	+79.77	+79.8	+83.7

Table 3.
Thermodynamic analysis of formation reactions C_3A and C_4AF according to data of [12] and formation reaction $C_4A_3\overline{S}$ according to data of [13].

Free lime begins to accumulate in the clinker from the decomposition beginning temperature of calcium carbonate to the appearance of the liquid phase due to the difference in the lime content in the tricalcium aluminate C_3A accepted in calculation and actually formed calcium monoaluminate CA. Totally because of gypsum decomposition and the difference between the limes content in the calcium aluminates, well-crystallized free lime in an amount of about 5% can be formed in the clinker from the decomposition beginning temperature of calcium carbonate until the appearance of the liquid phase, and this amount is sufficient to suppress the alite formation when the liquid phase appears.

Since the alite formation in clinker is suppressed and only belite is formed, so due to the difference in the lime content in these minerals the total content of free lime at the synthesis completion temperatures is already about 11%. When the melt temperature rises up to 1600 °C a small amount of alite is formed but the CaO_{free} however does not dissolve and is remained in an amount of about 6%.

To prevent the formation of CaO_{free} in high-sulfate clinker it is proposed to calculate the raw mix of Portland cement clinker in accordance with our patent [15] in two stages. At the first stage the calculation of calcium monoaluminate $CaO{\cdot}Al_2O_3$ synthesis in a high-sulphate clinker is made, meanwhile during burning intermediate metastable phase – ye'elimite $C_4A_3\,\overline{S}$ will form in a high-sulphate clinker. It decays when the liquid phase appears.

Since the formation of high-base phases is thermodynamically more likely when a liquid phase occurs, and C_3A and C_4AF can only be formed from low-base phases if free lime is present by reactions:

$$CaO \cdot Al_2O_3 + 2CaO \rightarrow 3CaO \cdot Al_2O_3, \quad (3)$$

$$CaO \cdot Al_2O_3 + CaO \cdot Fe_2O_3 + 2CaO \rightarrow 4CaO \cdot Al_2O_3 \cdot Fe_2O_3. \quad (4)$$

At a burning temperature of about 1300°C in the absence of CaO_{free} its source can only be the reaction of converting alite to belite and the decomposition of calcium sulfate by reactions:

$$3CaO \cdot SiO_2 \rightarrow 2CaO \cdot SiO_2 + CaO, \quad (5)$$

$$2CaSO_4 \rightarrow 2CaO + 2SO_2\uparrow + O_2\uparrow. \quad (6)$$

Since it is possible to convert alite to belite in a high-sulphate clinker by reaction 5, the calculation of the raw mix at the first stage is made for the formation of the maximum amount of alite in it which is possible at SC = 1. The calculation of the saturation coefficient of the raw mix by lime at the first stage is made using the well-known formula of Kinda V.A. [16] with SC = 1 for the formation of CA in the clinker (the coefficient for Al_2O_3 is 0.55):

$$SC = \frac{CaO - 0,55Al_2O_3 - 0,35Fe_2O_3 - 0,7SO_3}{2,8SiO_2}. \quad (7)$$

At a conclusion of this formula are used molar parities CaO, Al_2O_3, Fe_2O_3 and SiO_2 at formation in clinker the basic clinker minerals C_3S, C_2S, C_3A and C_4AF.

English analogue of the formula given is the formula for calculation LSF [17]:

$$LSF = \frac{CaO}{2,8SiO_2 + 1,2Al_2O_3 + 0,65Fe_2O_3}. \quad (8)$$

Factors in the given formula are taken from phase diagram CaO-Al_2O_3-SiO_2 and CaO-Al_2O_3-SiO_2-Fe_2O_3 at optimum relationship oxides providing absence free lime in clinker. If to use Kinda V.A's design procedure [16], that the formula (8) becomes full analogue of the formula (7):

$$LSF = \frac{CaO}{2,8SiO_2 + 1,65Al_2O_3 + 0,35Fe_2O_3 + 0,7SO_3}. \tag{9}$$

For calculation degree of saturation (DS) of clinker metastable minerals by sulphate in the presence of gypsum Atakuziev T.A.'s formula [18] is used considering that sulfatization C_2S, CA in clinker is possible. Taking into account the given updating the formula DS calculation looks as follows:

$$DS = \frac{SO_3 - 0,261Al_2O_3}{0,667SiO_2}. \tag{10}$$

At DS = 0 $C_3A_3C\bar{S}$ will be formed in the raw mix and at DS = 1 sulfosporite $2(C_2S)C\bar{S}$ will also be formed.

At the second stage the calculation of the synthesis of alite Portland cement with the required modular characteristics is made. It is assumed that when the liquid phase appears the gypsum is completely decomposed by the reaction 6 and the chemical composition of the clinker formed after gypsum decomposition is considered to be the chemical composition of one of the raw mix components. Other components of the raw mix are limestone and a corrective additive – quartz sand. At the second stage the saturation coefficient (LSF) is calculated using the formula (9) but only for the synthesis of tricalcium aluminate $3CaO{\cdot}Al_2O_3$ in the clinker (the coefficient for Al_2O_3 is 1.65).

4. Example of calculation of high-sulphate raw material mixture No. 1

At the first stage on the basis of raw components the chemical composition of which is shown in **Table 1** the raw mix of high-sulphate clinker is calculated for the synthesis of calcium monoaluminate in it according to the formulas (9) and (10) with the modular characteristics LSF = 1 and DS = 0. At DS = 0 calcium sulfoaluminate $C_3A_3C\bar{S}$ can be formed in the raw mix based on calcium monoaluminate.

Table 4 shows the calculated composition of the raw mix and the chemical composition of clinkers before and after the gypsum decomposition.

At the second stage a typical Portland cement clinker with modular characteristics LSF = 0.92, n = 2.3, p = 1.7 is calculated on the basis of clinker after decomposition of gypsum as one of the components of the raw mix and corrective additives: limestone and quartz sand.

The calculated composition of the raw mix for obtaining Portland cement clinker and it's chemical composition is shown in **Table 5**.

Finally the composition of the raw mix is calculated by multiplying the quantity of raw components of the clinker shown in **Table 4** by the quantity of clinker shown in **Table 5**, and is summed up with the quantity of raw components shown in **Table 5**:

$$CaCO_3 = (71.31x0.633) + 30.7 = 75.8\%; \tag{11}$$

$$Clay = 25.71x0.633 = 16.3\%; \tag{12}$$

Clinker	Raw mix composition			Chemical composition of clinker, mass %					
	Lime-stone	Clay	Gyp-sum	CaO	SiO_2	Al_2O_3	Fe_2O_3	SO_3	Total
Clinkers before the gypsum decomposition	71.31	25.71	2.98	64.66	20.39	8.06	4.79	2.1	100
Clinkers after the gypsum decomposition	71.31	25.71	2.98	66.05	20.83	8.23	4.89	0	100

Table 4.
The calculated composition of the raw mix and the chemical composition of clinkers before and after the gypsum decomposition.

Clinker	Raw mix composition			Chemical composition of clinker, mass %					
	Lime-stone	The first stage clinker	SiO_2	CaO	SiO_2	Al_2O_3	Fe_2O_3	SO_3	Total
Portland cement clinker	30.7	63.3	6.0	68.2	22.1	6.1	3.6	0	100

Table 5.
The calculated composition of the raw mix for obtaining alite Portland cement clinker and its chemical composition.

$$SiO_2 = 6.0\%; \tag{13}$$

$$\text{Gypsum} = 2,98x0,633 = 1,97\%. \tag{14}$$

The actual gypsum content after the introduction of corrective additives will decrease by:

$$\text{Gypsum} = 2.98 - 1,97 = 1.01\%. \tag{15}$$

Then the quantity of corrective additives is calculated. The quantity of corrective additives is equal to the difference between the quantity of raw components calculated using the formulas (11)–(14) and the quantity of raw components calculated at the first stage and shown in **Table 4**.

The quantity of corrective additives is equal to:

$$CaCO_3 = 75.8 - 71.31 = 4.49\%; \tag{16}$$

$$\text{Clay} = 16.3 - 25.71 = -9.4\%; \tag{17}$$

$$SiO_2 = 6.0 - 0 = 6.0\%. \tag{18}$$

A negative value for the clay amount means that it should be reduced.

Figure 3 shows the data of qualitative phase analysis of clinker prepared in accordance with correction calculation No. 1 and burnt at temperatures of 1150, 1200, 1250 and 1300°C.

X-ray analysis indicates the absence of CaO_{free} in a clinker prepared in accordance with corrective calculations No. 1, at a burning temperature of 1200°C or higher. Stable alite in such clinker is formed already at the burning temperature of 1250°C as evidenced by the appearance of diffraction maxima with d = 1.76 Å and d = 3.04 Å which are typical for alite at this temperature. The change in the intensity of the diffraction maxima of the main phases of the clinker prepared in accordance with the correction calculation No. 1 and burnt at temperatures of 1150, 1200, 1250 and 1300°C is shown in **Figure 4**.

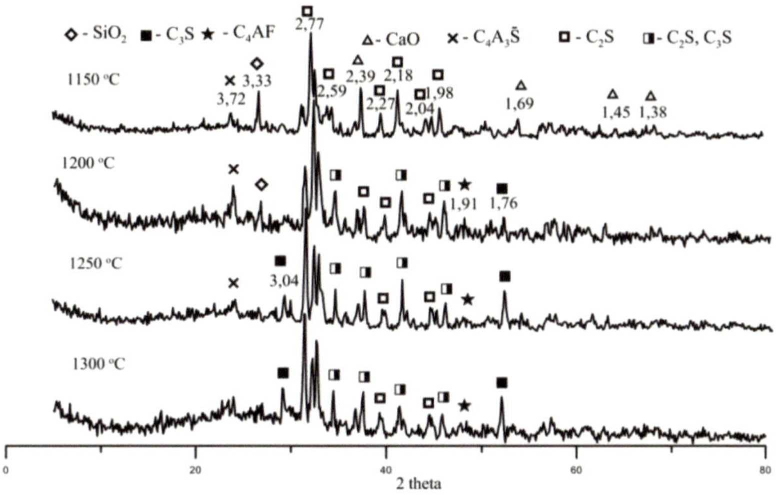

Figure 3.
The data of qualitative phase analysis of clinker prepared in accordance with correction calculation No. 1 and burnt at temperatures of 1150, 1200, 1250 and 1300°C.

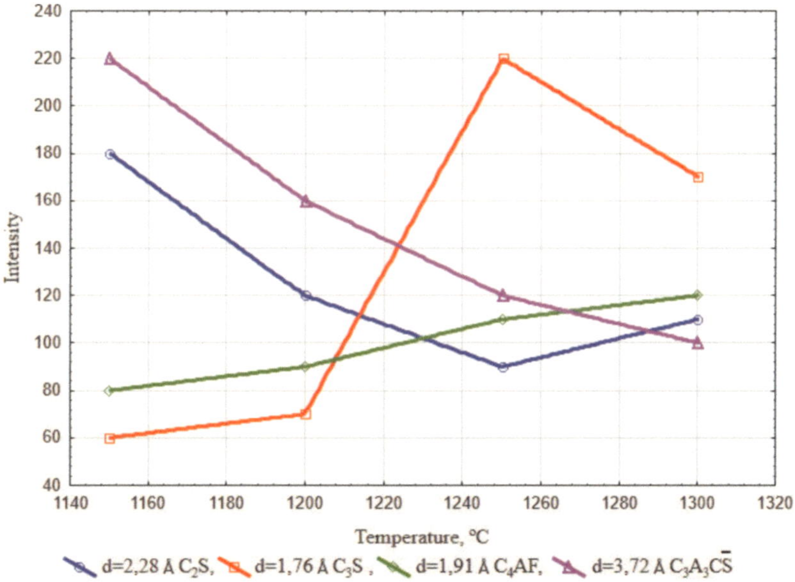

Figure 4.
The change in the intensity of the diffraction maxima of the main phases of the clinker prepared in accordance with the correction calculation No. 1 and burnt at temperatures of 1150, 1200, 1250 and 1300°C.

Diffraction peak with d = 1.76 Å which is typical for C_3S, increases up to a temperature of 1250°C and decreases starting from a temperature of 1250°C. The diffraction peak with d = 2.28 Å which is typical for C_2S on the contrary decreases to a temperature of 1250°C and increases at temperatures above of 1250°C which indicates the transformation of alite part into belite according to Eq. (5). The intensity of the diffraction maximum with d = 3.72 Å which is typical for $C_3A_3C\bar{S}$, decreases above the temperature of 1150°C, which indicates the decomposition of $CaSO_4$ in accordance with the reaction (6) and decay as a result.

To determine the actual phase composition of the clinker prepared in accordance with the correction calculation No. 1 and synthesized at a temperature of 1300°C a quantitative x-ray phase analysis was performed and it is shown in **Figure 5**.

Table 6 shows the phase composition of clinker synthesized in accordance with correction calculation No. 1 based on quantitative phase analysis.

On the basis of clinker prepared in accordance with corrective calculation No. 1 and synthesized at a temperature of 1300°C Portland cement was prepared by joint grinding of clinker with gypsum dihydrate (**Figure 6**). The cement activity was

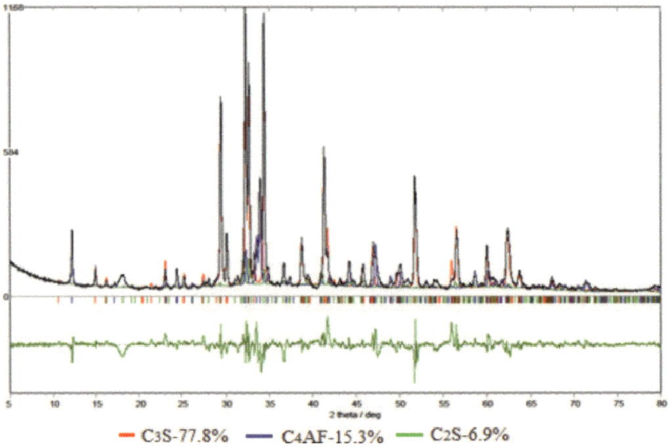

Figure 5.
The phase analysis data of clinker synthesized on the basis of raw mix in accordance with correction calculation No. 1.

Name of mineral phase	Guantity in clinker, mas. %
Three calcium silicate (alite) C_3S	77.8
Two calcium silicate (belit) C_2S	6.9
Brownmillerit C_4AF	15.3

Table 6.
The phase composition of clinker synthesized in accordance with correction calculation No. 1.

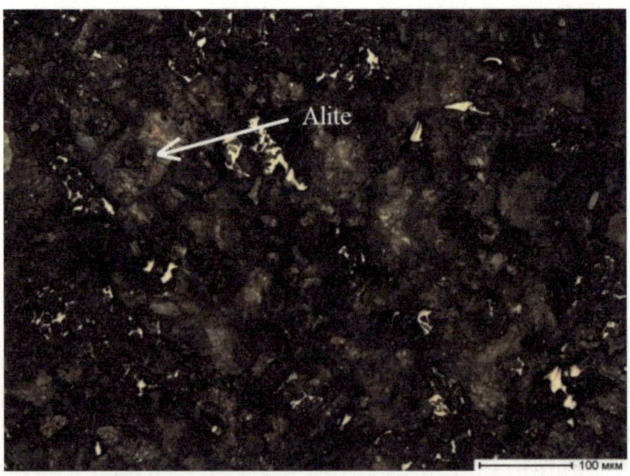

Figure 6.
A micrograph of clinker synthesized in accordance with correction calculation No. 1.

Clinker type	Composition, %		SO_3, %	S*, m2/kg	R008, %	ND, %	Setting time, hour-minute	
	Clinker	Gypsum					Initial	Final
Clinker No. 1	96	4	3.96	348	12.7	25.2	2–50	4–15

**S – Blaine's specific surface; R008 – residue on the sieve No. 008; ND – normal density.*

Table 7.
The physical and mechanical properties of Portland cement.

Clinker type	Compressive strength, MPa, after, days			
	2	7	14	28
Clinker No. 1	21.8	31.1	42.5	67.3

Table 8.
Portland cement compressive strength.

determined on cubes with a size of 2x2x2 cm prepared from cement paste of normal density. The physical and mechanical properties of cement are shown in **Tables** 7 and **8**.

5. Example of calculation of high-sulphate raw material mixture No. 2

At the first stage on the basis of raw components the chemical composition of which is shown in **Table 2** the raw mix of high-sulphate clinker is calculated for the synthesis of calcium monoaluminate in it according to the formulas (9) and (10) with the modular characteristics LSF = 1 and DS = 1. At DS = 1 calcium sulfoaluminate $C_3A_3C\overline{S}$ can be formed in the raw mix based on calcium monoaluminate and sulfospurrit $2(C_2S)C\overline{S}$ can be formed in the raw mix based on belite.

Table 9 shows the calculated composition of the raw mix and the chemical composition of clinkers before and after the gypsum decomposition.

At the second stage a alite Portland cement clinker with modular characteristics LSF = 0.92, n = 2.3, p = 1.7 is calculated on the basis of clinker after decomposition of gypsum as one of the components of the raw mix and corrective additives: limestone and quartz sand.

The calculated composition of the raw mix for obtaining alite Portland cement clinker and its chemical composition is shown in **Table 10**.

Finally the composition of the raw mix is calculated by multiplying the quantity of raw components of the clinker shown in **Table 9** by the quantity of clinker

Clinker	Raw mix composition			Chemical composition of clinker, mass %					
	Lime-stone	Clay	Gypsum	CaO	SiO_2	Al_2O_3	Fe_2O_3	SO_3	Total
Clinkers before the gypsum decomposition	59.8	21.3	18.9	60.3	16.5	6.6	3.9	12.7	100
Clinkers after the gypsum decomposition	59.8	21.3	18.0	69.1	18.9	7.5	4.5	0	100

Table 9.
The calculated composition of the raw mix and the chemical composition of clinkers before and after the gypsum decomposition.

Clinker	Raw mix composition			Chemical composition of clinker, mass %						
	Lime-stone	**The first stage clinker**	**SiO_2**	**CaO**	**SiO_2**	**Al_2O_3**	**Fe_2O_3**	**SO_3**	**$\sum$**
Portland cement clinker	20.8	72.9	6.4	68.3	22.1	6.0	3.6	0	100

Table 10.
The calculated composition of the raw mix for obtaining alite Portland cement clinker and its chemical composition.

shown in **Table 10**, and is summed up with the quantity of raw components shown in **Table 10**:

$$CaCO_3 = (59,8x0,729) + 20,8 = 64,3\%; \quad (19)$$

$$\text{Clay} = 21,3x0,729 = 15,5\%; \quad (20)$$

$$SiO_2 = 6,4\%; \quad (21)$$

$$\text{Gypsum} = (18,9x0,729) = 13,8\%. \quad (22)$$

Then the quantity of corrective additives is calculated. The quantity of corrective additives is equal to the difference between the quantity of raw components calculated using the formulas (19) and (20) and the quantity of raw components calculated at the first stage and shown in **Table 9**.

The quantity of corrective additives is equal to:

$$CaCO_3 = 64,3 - 59,8, = 4,4\%; \quad (23)$$

$$\text{Clay} = 15,5 - 21,3 = -5,8\%; \quad (24)$$

$$SiO_2 = 6,4 - 0 = 6,4\%; \quad (25)$$

A negative value for the clay amount means that it should be reduced the same as in the calculation example No. 1.

The actual gypsum content after the introduction of corrective additives will decrease by:

$$CaSO_4 = 18,9 - 13,8 = 5,1. \quad (26)$$

Figure 7 shows the data of qualitative x-ray phase analysis of clinker prepared in accordance with calculation mentioned above and burnt at temperatures of 1100, 1200, 1300 and 1350°C.

X-ray analysis indicates the absence of CaO_{free} in a clinker prepared in accordance with corrective calculations No. 2 at a burning temperature above 1300°C. Stable alite C_3S is formed at the burning temperature of 1350°C as evidenced by the appearance of diffraction maxima with d = 1.76 Å and d = 3.04 Å which are typical for alite at this temperature.

The intensity of the diffraction maximum with d = 3.72 Å which is typical for $C_3A_3C\,\overline{S}$ increases to a temperature of 1300°C but above this temperature it is not fixed which indicates the decomposition of $CaSO_4$ in accordance with the reaction (No. 6) and decay as a result.

According to the qualitative phase analysis data a significant amount of gypsum is released up to the burning temperature of 1300°C which is fixed by the diffraction maximum with d = 3.47 Å. Gypsum remains are fixed even at a temperature of 1350°C.

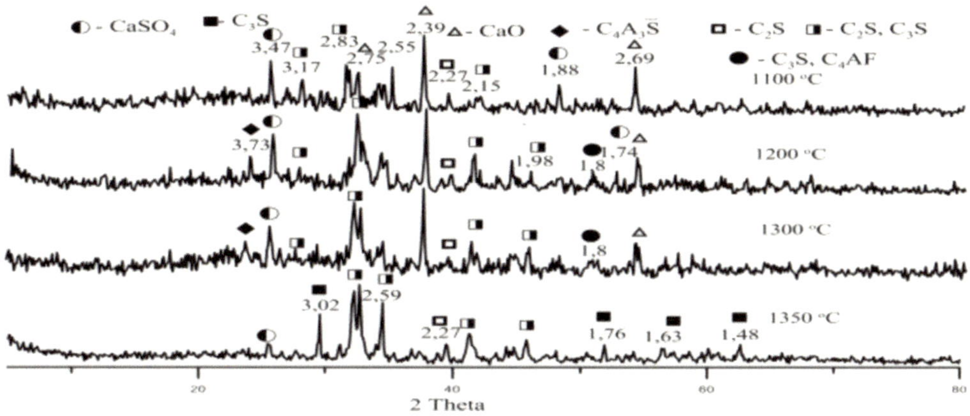

Figure 7.
The data of qualitative phase analysis of clinker prepared in accordance with calculation No. 2 and burnt at temperatures of 1100, 1200, 1300 and 1350°C.

To determine the actual phase composition of the clinker prepared in accordance with the correction calculation No. 2 and synthesized at a temperature of 1350°C a quantitative x-ray phase analysis was performed. Its results shown in **Figure 8**.

Table 11 summarizes the results of quantitative x-ray phase analysis of synthesized clinker.

The test results show that in accordance with corrective calculation No. 2 a significant amount of C_3S is retained when preparing clinker based on a high-sulphate raw mix that initially contains 12.7% SO_3 (**Figure 9**).

On the basis of clinker prepared at temperature of 1350°C Portland cement was prepared by joint grinding of clinker with natural gypsum. The cement activity was determined on cubes with a size of 2x2x2 cm prepared from cement paste of normal density.

The physical and mechanical properties of cement are shown in **Tables 12** and **13**.

Examples data of corrective calculations show that using the proposed calculation method it is possible to save a significant amount of C_3S in a clinker synthesized on the basis of a high-sulphate raw mix.

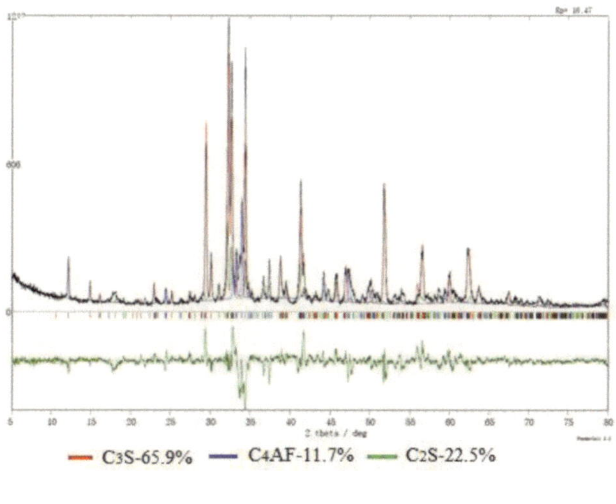

Figure 8.
The phase analysis data of clinker synthesized on the basis of raw mix in accordance with correction calculation No. 2.

Name of mineral phase	Guantity in clinker, mas. %
Three calcium silicate (alite) C_3S	65.9
Two calcium silicate (belit) C_2S	22.5
Brownmillerit C_4AF	11.7

Table 11.
The phase composition of clinker synthesized in accordance with correction calculation No. 2.

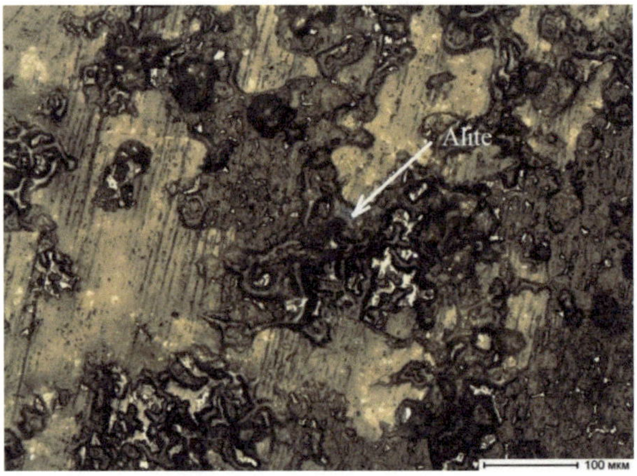

Figure 9.
A micrograph of clinker synthesized in accordance with correction calculation No. 2.

Clinker type	Composition, %		SO_3, %	S*, m2/kg	R008, %	ND, %	Setting time, hour-minute	
	Clinker	Gypsum					Initial	Final
Clinker No. 2	100	0	12.7	378	3.1	28.3	1–50	2–15

**S – Blaine's specific surface; R008 – residue on the sieve No. 008; ND – normal density.*

Table 12.
The physical and mechanical properties of Portland cement.

Clinker type	Compressive strength, MPa, after, days			
	2	7	14	28
Clinker No. 2	10.1	14.7	20.0	44.7

Table 13.
Portland cement compressive strength.

The quantity of corrective additives depends on the modular characteristics of the synthesized clinker. When limiting the modular characteristics of Portland cement clinker LSF = 0.92–0.98; n = 2.0–3.0; p = 1.7–4.0 the minimum amount of additives equal to 4.0% is introduced with the minimum values of modular characteristics, i.e. LSF = 0.92; n = 2.0; p = 1.7; and with the minimum amount of gypsum introduced, i.e. DS =0. The maximum amount of corrective additives equal to 23.0% is introduced with the maximum values of modular characteristics, i.e. LSF = 0.98; n = 3.0; p = 4.0 and with the maximum amount of gypsum introduced, i.e. DS =1.

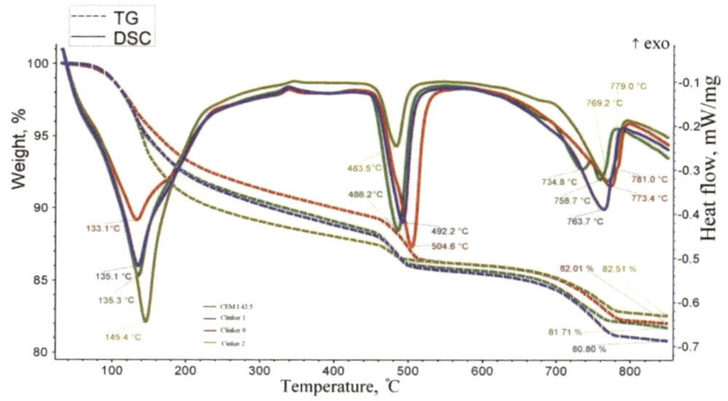

Figure 10.
Results of comparative thermal analysis (TA) of hydration products.

Clinker type	The value of the endo effect, J/g	Mass loss, %
Clinker 0	148.2	17.9
Clinker 1	167.0	19.2
Clinker 2	180.5	17.5
CEM I 42.5	155.0	18.3

Table 14.
Values of the first endo-effect and the total mass loss of samples.

The quantity of clay removed from the raw mix correlates with the amount of mix additives put in the raw mix, i.e. with the same modular characteristics of the clinker the minimum quantity of clay removed corresponds to a minimum quantity of additives and the maximum quantity of clay removed corresponds to the maximum quantity of additives.

At the final stage of investigation comparative thermal analysis (TA) of hydration products was conducted. Results of tests are resulted on **Figure 10**.

The **Table 14** shows the values of the first endo-effect and the total mass loss of samples.

At hydration alite attaches 5 molecules of water, but belite only 2. Due to this difference, the clinker 0 has smallest first endo effect.

6. Conclusion

The present study revealed that the SO_3 negative impact on the Portland cement clinker synthesis, resulted in a C_3S content reduction and the C_2S and C3A content increasing in the final product. It leads to lowering clinker fire resistance and cement quality due to the thermodynamic preference of the ye'elimite $C_4A_3\bar{S}$ synthesis in the presence SO_3 and, consequently, the presence of low-basic calcium monoaluminate CA in the synthesized clinker. In the presence of calcium monoaluminate, solid-phase synthesis of high-base C_3A and C_4AF is thermodynamically impossible. As a result, free lime accumulates in the synthesized clinker, which prevents the liquid-phase synthesis of C_3S.

A method for elimination of the SO_3 negative impact on the Portland cement quality by calculating the raw material mixture composition with a significant amount of SO_3 has been developed and patented.

Abbreviations

The following abbreviations are used in this manuscript:

C_3S	$3CaO \cdot SiO_2$;
C_2S	$2CaO \cdot SiO_2$;
C_3A	$3CaO \cdot Al_2O_3$;
C_4AF	$4CaO \cdot Al_2O_3 \cdot Fe_2O_3$;
$C_3A_3C\bar{S}$	$3CaO \cdot 3Al_2O_3 \cdot CaSO_4$;
$2(C_2S)C\bar{S}$	$2(2CaO \cdot SiO_2) \cdot CaSO_4$;
$C_{12}A_7$	$12CaO \cdot 7Al_2O_3$;
CaO_{free}	free CaO;
SC	Saturation Coefficient;
LSF	Lime Saturation Factor;
DS	Degree of Saturation by Sulfate;
LOI	Loss On Ignition.

Author details

Oleg Sheshukov* and Michael Mikheenkov
Ural Federal University Named after the First President of Russia B.N. Yeltsin, Institute of Metallurgy of the Ural Branch of the Russian Academy of Sciences, Ekaterinburg, Russian Federation

*Address all correspondence to: o.j.sheshukov@urfu.ru

References

[1] Factors el Mar Cortada Mut, Linda Kaare Norskov, Peter Glarborg, Kim Dam-Johansen Sulphur release from alternative fuel firing / Maria del Mar Cortada Mut & other//Global Cement Magazine 9 (2014) p. 36–41.

[2] Július Strigáč Effect of Selected Alternative Fuels and Raw Materials on the Cement Clinker Quality/ Strigáč Július // Journal Of Civil Engineering Vol. 10, Issue 2, 2015 p. 81–92.

[3] Ying-Liang Chen, Juu-En Chang , Ming-Sheng Ko Reusing Desulfurization Slag in Cement Clinker Production and the Influence on the Formation of Clinker Phases/ Chen Ying-Liang & other// Sustainability 9, 1585 (2017) p.1–14.

[4] Javed I. Bhatty.; Role of Minor Elements in Cement Manufacture and Use, Portland Cement Association, Skokie, Illinois 1995, p.24–25.

[5] M. Yamashita, H. Tanaka Low Temperature Burnt Portland Cement Clinker Using Mineralizer/ Yamashita M., Tanaka H. // Cement Science and Concrete Technology № 65 (2011) p. 82–87.

[6] H.F.W. Taylor, Cement Chemistry, Academic Press, London, 1990.

[7] Sayed Horcoss Influence of the clinker SO_3 on the cement characteristics / Horcoss Sayed, Ltief Roger, RizkToufic//Cement And Concrete Research 41 (2011). p. 913–919. Available from: http://www.sciencedirect.com/science/article/pii/S0008884611001268

[8] Mieke De Schepper, Philip Van den Heede, Eleni C. Arvaniti, Klaartje De Buysser Sulfates in Completely Recyclable Concrete and the effect of $CaSO_4$ on the clinker mineralogy/ Schepper De & other //Construction and Building Materials 137 (2017) p. 300–306.

[9] Sayed Horkoss, Roger Lteif, and Toufic Rizk Calculation of the C_3A Percentage in High Sulfur Clinker / Horcoss Sayed & other // International Journal of Analytical Chemistry Volume 2010, Article ID 102146, p. 1–5.

[10] Laure Pelletier-Chaignat, Frank Winnefeld, Barbara Lothenbach, Gwenn Le Saout Influence of the calcium sulphate source on the hydration mechanism of Portland cement–calcium sulphoaluminate clinker–calcium sulphate binders/ Pelletier-Chaignat Laure & other // Cement & Concrete Composites 33 (2011) p. 551–561.

[11] Donald H. Campbell. Microscopical Examination and Interpretation of Portland Cement and Clinker Second Edition/ Published by:Portland Cement Association – USA, 1999. 214 p.

[12] Babushkin, V.I., Matveev G.M., Mchedlov-Petrosjan, O.M. Thermodynamics Of Silicates - M: Stroyizdat, 1986.408 p.

[13] Kuznezova T.V. Aluminate and sulfoaluminate cements/ - M.: Stroyizdat, 1986., 208 p.

[14] Michael A. Miheenkov Compacting as a way of reception sulfated hydraulic binders/Vestnik MGSU.- – 2011. – № 1 . – p. 131–142

[15] Patent 2527430 Russian Federation: MPK C04B 7/36. Method for Correcting Composition of Portland Cement Clincer Based on Highe-Sulphate Crude Mixture. Author: Michael A. Miheenkov., Patent Holder: Michael A. Miheenkov. – 2013112990/03; Date of declare. 22.03.2013 ; Date of publication. 27.08.2014. Bulletin. № 24.

[16] Kind V.A. Chemical characteristics of Portland cement/ L.-M. Gosstroyizdat, 1932..

[17] Walter H. Duda Cement.; Data-Book /Internationale Verfahrenstechniken der Zementindustrie 2., Auflage Bauverlag GmbH Wfmbaden and Berlin, 1981, 484 p.

[18] Atakuziev, T.A. Sulfomineral cements on a basis of phosphogypsum/ T.A.Atakuziev, F.M.Mirzoev.: "FAN", 1979. – 152 p.

Section 3

Comparative Parameters of Cement Production

Chapter 7

Energy and Economic Comparison of Different Fuels in Cement Production

Oluwafemi M. Fadayini, Clement Madu, Taiwo T. Oshin, Adekunle A. Obisanya, Gloria O. Ajiboye, Tajudeen O. Ipaye, Taiwo O. Rabiu, Joseph T. Akintola, Shola J. Ajayi and Nkechi A. Kingsley

Abstract

Cement clinkerisation is the major energy-consuming process in cement manufacturing due to the high-temperature requirement. In this paper, energy data including specific energy consumption, forms, and types of energy used at different units of cement manufacturing processes were analyzed and compared for effectiveness, availability, cost, environmental, and health impact. Data from three different cement industries in Nigeria labeled as A, B, and C were used for the analysis in this study. The results of this research work established that coal is the cheapest energy source but environmental issues exonerate it from being the choice energy source. LPFO and Natural gas give better production output while minimizing pollution and health issues. When benchmarked against each other, Factory B was found to be the most energy-efficient in terms of output and cost of production. Although coal is cheaper compared to fuel oil and supposed to contribute a share of fuel used in cement industries, the industries are moving towards the use of alternative and conventional fuels to reduce environmental pollution. It is therefore recommended that deliberate effort to achieve appreciable energy-efficient levels should be the priorities of the cement industries in Nigeria.

Keywords: Cement, Coal, Fuel oil, Natural gas, Energy Consumption, Energy source, Clinkerization

1. Introduction

Cement is regarded as a binder, a material useful in building and civil construction that hardens and adheres to other substances to bind them together. Cement is rarely used only, but to bind other building materials such as gravel and sand together. When mixed with fine aggregates, it is used to produce mortar for mansory or with gravel and sand, it produces concrete. Energy consumption in the Industrial sector ranges from 30–70% of the total energy used in some selected countries as previously reported by [1]. The cement sub-sector utilizes nearly 12–15% of entire industrial energy usage [1, 2] due to the high temperatures required in the kilns. Cement is a vital product used in society for constructing

IntechOpen

modern infrastructure as well as safe and comfortable buildings. Cement manufacturing is an energy-intensive process due to the high temperature required in the kilns for clinkerization. Energy cost contributes to about 40–50% of cement production cost in Nigeria depending on the production process and type of cement with 1 tonne of cement requiring 60–130 kg of fuel or its equivalent and about 105 kWh of electricity [3].

Fossil fuels like coal, pet coke, fuel oil, and gas are the primary fuels used in the cement kilns. These fuels which exist in solid, gaseous, and liquid also provide most of the global energy needs and demand. Some of these fuels e.g. coal and natural gas are utilized in their natural form while energy resources like petroleum, shale, and bituminous sands require processing, refinement, and distillation to produce consumable fuels.

The conservation of energy is an essential step to take towards overcoming the mounting problems of the worldwide energy crisis and environmental degradation. In particular, developing countries are interested in increasing their awareness of the energy efficiency in power generation and consumption in their countries. However, usually, only limited information/sources on the rational usage of energy are available [4].

The energy source or mix to be implemented will have to meet the varying energy demand of the countries, industry, or organizations as well as improving the security against the energy crisis. Fuel availability, ease of processing and handling, environmental pollution, storage, and cost are some of the factors that determine the selection of fuel [5].

In cement production, the energy use is distributed as 92.7% for pyro-processing, 5.4% for finishing grinding, and 1.9% for raw grinding [6]. The type of fuel used determines the quantity of greenhouse gases (GHG) emission, cement product quality, and cost. Large volumes of CO_2 are emitted during cement production and it is believed that this sector represents 5%–7% of the total CO_2 anthropogenic emissions [7, 8]. Environmental concerns are of great importance since cement and the production of its raw materials are extensively based on fossil fuels.

There are three processes in cement manufacturing plant [9]: raw material mixing, pyroprocessing (burning), and grinding.

Raw material processing: this can be the wet process or dry process depending on the method of milling. In the wet process, raw materials other than plaster are crushed to a diameter of approximately 20 mm by a crusher and mixed in an appropriate ratio using an automatic weigh balance. Its particle size is further reduced to finer particles by tube mill of 2–3.5 m diameter and length 10–14 m in the presence of water from a slurry of 35–40% [10]. In the dry process, the raw materials (calcareous and argillaceous) are separately crushed to about 2–5 cm. They are later dried in a cylindrical rotary drier having a diameter of 2 m and a length of about 20 m, pulverized into fine particles, and stored. The pulverized fine raw materials are then mixed automatically in proportions to form a uniform dry mix and sent to a kiln for clinker production where about 80% of the energy used in cement production is consumed [4, 11]. The electrical energy requirement of the dry process is higher compared to the wet process while the thermal energy consumption is very low compared to the wet process. The primary energy consumption in a typical dry process is about 75% fossil fuel and up to 25% electrical energy [1].

The pyroprocessing in the kiln generates about 81% of cement production CO_2 emission; 36.8% from fuel combustion while 46.3% is from pyroprocessing reaction [6]. Hence, the choice of fuel and energy conversion efficiency have a net effect on cement CO_2 emission. The exact consumption of energy in the production of cement varies from one technological approach to another." The major fuel used in clinker production is coal and petroleum coke but alternate energy source like biomass, waste heat, fuel oil, solvent, tyres, gas, etc. are becoming attractive in recent years [12]. A considerable amount of energy is consumed in manufacturing cement. Thus, the focus should be centered on energy savings and energy-associated environmental emissions both locally and universally [13–17]. The chief share of the total thermal energy consumption is required by pyro-processing and it accounts for approximately 93–99% of the entire fuel consumption [1]. Though electrical energy is principally used for the operation of the raw materials which accounts for 33% of its consumption, and clinker crushing and grinding equipment which accounts for 38% of its consumption. Electrical energy is needed to operate equipment such as combustion air blowers, kiln motors, and fuel supply, etc. accounting for 22% of its consumption to sustain the pyro-process.

The calorific value of common fuels used in cement production is shown in **Table 1**. Natural gas has the highest energy content followed by fuel oil while coal has the least energy content of the three fuels.

Coal is regarded as the most abundant fossil fuel on earth, with a global recoverable reserve estimated at 216 years [18]. Coal provides 26% of global primary energy consumption and contributes 41% of global electricity generation.

Fuel oil is a distillate or a residue fraction produced from petroleum distillation. It is any liquid from petroleum that is burned in a furnace for heat and power generation In terms of industrial use of fuel especially in cement kiln firing, heavy fuel oil, or low pour fuel oil (LPFO). Heavy oil is a long residue obtained from the atmospheric distillation column. Heavy fuel oil is used mainly to produce electricity, to fire boiler and furnace in industry, notable the cement, pulp, and paper, and to power large marine and other vessels.

Natural gas is a fossil fuel like oil and coal, thus it is essentially the remains of plants, animals, algae, and microbes that lived millions of years ago. Over the years, natural gas has secured its vital role in every aspect of world development, particularly its role to replace coal and oil with having a high energy content that the two aforementioned.

In Nigeria, cement production has increased exponentially from 2 million tonnes in 2002 to about 17 million in 2011 [19]. Thus, making Nigeria's cement industry contributing about 60% of the West African region's cement output in 2011. Since the sector consumes a considerable amount of energy, it is necessary to identify and reduce energy wastage [20]. Also, the unit fuel cost for cement production in Nigeria is $30 per tonne which is very high compared to an advanced country like China ($6 per tonne) thereby contributing to the high cost of unit price cement [21]. The use of energy utilization analysis for energy and financial savings has generated research interest in recent years [22]. Therefore, this research work aims to analyse the cost vis-à-vis the pollution tendencies of each energy source and its consequence on health and the environment from the energy data obtained from the cement industries.

S/N	Fuel	Energy Content (MJ/Kg)
1	Coal	36.3
2	Natural gas	54.0
3	Fuel oil	45.6

Source: Engineering ToolBox, (2008). Fossil and Alternative Fuels Energy Content. *[online] Available at: https://www.engineeringtoolbox.com/fossil-fuels-energy-content-d_1298.html [Accessed: 23/2/2021].*

Table 1.
Energy contents of coal, fuel oil, and natural gas.

2. Methodology

2.1 Data collections

Three major Cement producers in Nigeria (Dangote Cement in Obajana, Kogi State; United Cement Company in Calabar, UNICEM - Cross River and Nigerian Cement Company in Nkalagu, NIGERCEM, Ebonyi State), labeled Factory A, B and C were approached for the Data on energies consumed during cement production and were collected for the analysis.

2.1.1 Calculations involved in the analysis

Specific heat:

The standard or universally accepted specific fuel consumption for clinker production is 720 Kcal/m^3 of clinker from:

$$\frac{\textit{Calorific value of gas} \times \textit{Total consumed}}{\textit{total clinker } x \ 1000} = \frac{720Kcal}{m^3} \tag{1}$$

Let n = 1.

From ideal gas law: PV = nRT

$$R = \frac{P_{act} \times V_{act}}{T_{act}} \ \textit{at a given temperature and pressure} \tag{2}$$

$$R = \frac{P_o \times V_o}{T_o} \ \textit{at normal condition } (0\ °C \textit{ and } 1.0 \ atm) \tag{3}$$

Therefore

$$\frac{P_{act} \times V_{act}}{T_{act}} = \frac{P_o \times V_o}{T_o} \tag{4}$$

$$\text{Volumetric flow rate } Q = \frac{V}{t} \tag{5}$$

Substituting Eq. (4) into (1) and simplifying we have

$$Q_o = Q_{act} \times \left[\frac{P_{act}}{P_o}\right] \times \left[\frac{T_o}{T_{act}}\right] \tag{6}$$

Where Q_o = Q_N = Gas flow rate at normal condition.

Q_{act} = measured flow rate in m^3/hr.

P_{act} = P measured + P ambient

P ambient = 1.01325 bar.

P_o = pressure at normal condition

T_o = temperature at normal condition

Therefore

$$Q_N = Q_{act} \times \left[\frac{P_{measured} \times P_{ambient}}{P_O}\right] \times \left[\frac{T_O}{T_{act}}\right]\left(\frac{Nm^3}{hr}\right) \tag{7}$$

Since we are dealing with volume and not flow rate. Then equation becomes this

$$V_o = V_{act} \times \left[\frac{T_o}{T_{act}}\right](Nm^3) \tag{8}$$

V_o = volume at normal condition.

Since all the pressure and temperature are in atmospheric and absolute units. In Eq. (8) the fluid in meter cubic (m^3) is converted to Normal cubic meter (Nm^3). The reversal of Eq. (8) converts the fluid in Normal cubic meter to meter cubic. Therefore Eq. (8) becomes:

$$V_{act} = V_o \times \left[\frac{P_o}{P_{act}}\right] \times \left[\frac{T_{act}}{T_o}\right](m^3) \tag{9}$$

V_{act} = measured volume in m^3.
n = number of moles.
R = molal gas constant.
T_{act} = temperature in Kelvin.
$P_{act} = P_{measured}$ + Pambient.

2.1.2 Cost analysis

The following cost of material for fuel oil, natural gas, and coal was as obtained [23–25], respectively;

Fuel oil (diesel) = ₦223.740 ($0.587) per litre.
Natural gas = $2.76 per 1000 ft^3.
Coal = $68.9 per tons.

The calculated cost in **Table 2** is subject to some conversions as the consumption of coal, fuel, and energy is given in tonnes. The cost for natural gas is given in $\$/ft^3$ and the cost of fuel oil is given in $ per liters. For the two cases where volume is used, the quantity consumed is converted from tonnes to ft^3 and liters for natural gas and fuel oil, respectively.

Density of fuel oil (diesel) = 0.85 kg/litre.
Density of natural gas = 0.68 kg/m^3.
Density of coal = 1506 kg/m^3.
1 tonnes = 1,000 kg.
1 m^3 = 35.315 ft^3

$$volume = \frac{mass}{density} \tag{10}$$

3. Results

3.1 Flue gas composition

Table 2 shows the various proportions of flue gases in coal, fuel oil, and natural gas (**Figure 1**).

4. Discussion of results

From the specific heat of consumption point of view, it is observed that of the three different cement companies that were used for analysis; the specific heat of

		COAL				FUEL				NATURAL GAS			
SN	Factory	Prod., tons	Total Fuel	C_P, Kcal/kg	Cost, $	Prod., tons	Total Fuel	C_P, Kcal/kg	Cost, $	Prod., tons	Total Fuel	C_P, Kcal/kg	Cost, $
1	A	6,000	720	600	49,608	6,014	428,755	697	2.96×10^8	6,081	470,470	698	67,435,924
		6,088	735	604	50,641.5	6,013	421,850	676	2.91×10^8	6,095	466,260	690	66,832,474
		6,148	745	606	51,330.5	6,074	425,987	686	2.94×10^8	6,168	460,000	673	65,935,182
		6,074	725	597	49,952.5	6,148	424,586	675	2.93×10^8	6,081	471,363	699	67,563,925
2	B	4,600	595	647	40,995.5	5,028	428,577	834	2.96×10^8	4,074	313,647	694	44,957,331
		4,720	608	644	41,891.2	5,145	412,580	784	2.85×10^8	3,563	310,840	787	44,554,983
		4,175	540	646	37,206	5,111	417,255	798	2.88×10^8	3,612	306,666	766	43,956,693
		4,834	621	642	42,786.9	5,273	435,674	808	3.01×10^8	3,554	314,242	798	45,042,616
3	C	5,125	612	597	42,166.8	5,412	385,789	697	2.65×10^8	3,798	282,283	670	40,461,698
		5,175	625	604	43,062.5	5,466	383,298	689	2.65×10^8	3,908	276,000	637	39,561,109
		5,164	616	596	42,442.5	5,533	382,037	675	2.64×10^8	3,807	279,756	662	40,099,485

Prod = 24 hours Production in tons; C_P = Specific heat, Kcal/kg; Total fuel = Total fuel consumed during the 24 hours of production, tonnes.

Table 2.
Calculated specific heat consumption and cost for the different fuel.

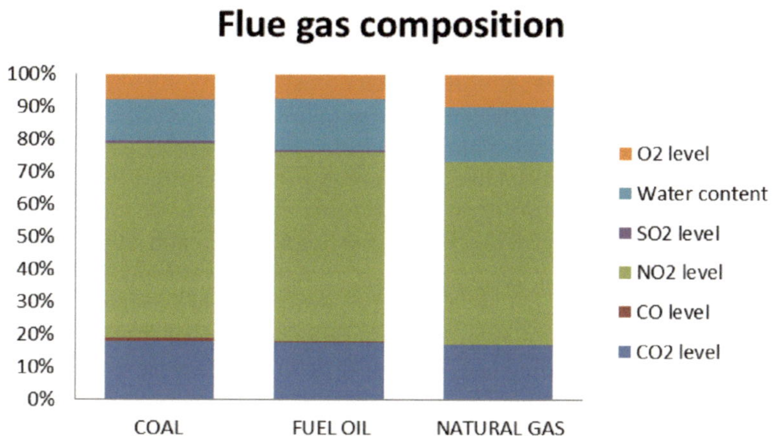

Figure 1.
Flue gas from Coal, Fuel oil, and Natural gas from gas analyses.

consumption of coal was less compared to that of fuel oil and natural gas (**Table 2**). This indicates that coal, as a good source of energy for firing in the clinker considering its high calorific value. On the other hand, the cost analysis revealed coal as the cheapest energy used by these cement companies as shown in **Table 2**, that is, 1 m^3 of coal was consumed at $103.766. For natural gas, 1 m^3 of it was consumed at $9.747 while 1 m^3 of LPFO was consumed at $586.8513.

Natural gas is the most readily available, and highly economical source of energy in use for the production of cement, compared to coal and fuel oil. Related results were reported by Ohunakin et al., [3] for Energy and Cost Analysis of Cement Production Using the Wet and Dry Processes in Nigeria. Based on the flue gases produced from these three sources of energy at Dangote cement (**Table 3**), Sulphur oxides emissions are relatively higher in coal and fuel oil than in natural gas. For carbon monoxide emission, this is high in coal followed by fuel oil while it is low in natural gas. Nitrogen oxide emissions are high in coal and fuel oil compared to natural gas. Also, Carbon Dioxide emission is high in coal and fuel oil compared to natural gas. In a similar study, Worrell et al., [26] report that fuels like coal and coke contribute to an increase in specific carbon dioxide emissions. Similarly, the Oxygen content is high in natural gas compared to coal and fuel oil. Based on the flue gases, natural gas presents itself as the most efficient and the most environmentally friendly source of energy.

DANGOTE	FLUE GAS (%) (From Gas Analyzer)		
Compound	**Coal**	**Fuel Oil**	**Natural Gas**
CO_2	18.01	17.73	17.02
CO	0.89	0.10	0.002
H_2O	12.80	15.51	16.71
NO_2	59.78	58.01	55.85
SO_2	0.82	0.92	0.05
CH_4	0.00	0.00	0.00
O_2	7.70	7.67	9.90

Table 3.
Comparison of flue gas from coal, fuel oil, and natural gas.

5. Conclusion

Although coal gave a cheaper consumption cost compared to fuel oil, for the production of cement, as expected, which could be used as an immediate substitute for natural gas, if peradventure its unavailability arises. Nevertheless, the environmental issues presented from its use as an energy source cannot be ignored. The LPFO (fuel oil) is quite expensive and would unavoidable impact on the cost of the final product. Benchmarking these three factories against each other, the cheapest energy consumption cost per ton production of cement was from Factory B while Factory A was the most expensive – for all three energy sources under investigation. From the analysis of the work, natural gas is one of the fossil fuels used in the production of cement. It is the cheapest amongst the three-fuel used in the production of cement and readily available. Also, natural gas emits lesser greenhouse gases to the environment, thereby lowering its effect on plant and animal health. Coal which is a close substitute is unavailable due to the closing down of Nigeria's coal mine and it poses too much threat to the environment and the health of plants and animals. Fuel oil is also available but as of now it is the most expensive fuel used in Nigerian cement industries and it also poses a high threat to the environment and life.

6. Recommendations

Energy sources have a direct impact on the market price of cement, the environment, and human health. Natural gas is an available energy source in Nigeria, and more economical and environmentally friendly compared to coal and fuel oil. It is therefore recommended - to cut down energy costs, guarantee power supply to the power plant, and minimize the emission of threats caused by cement industries to the environment. Factory B was most energy-efficient and a closer understanding of their process should be considered by Factory A and C. The unit cost of fuel oil component, the commonly used energy source in cement production in Nigeria is very high, over $15 as against $6 in China [3]. This is responsible for the high cost of cement in Nigeria. Thus, the need for an energy-efficient production process is recommended.

Conflict of interest

There is no conflict of interest associated with this work.

Author details

Oluwafemi M. Fadayini[1]*, Clement Madu[1], Taiwo T. Oshin[2], Adekunle A. Obisanya[3], Gloria O. Ajiboye[2], Tajudeen O. Ipaye[3], Taiwo O. Rabiu[4], Joseph T. Akintola[1], Shola J. Ajayi[1] and Nkechi A. Kingsley[1]

1 Department of Chemical Engineering, Lagos State Polytechnic, Ikorodu, Lagos, Nigeria

2 Department of Chemical Science, School of Pure and Applied Sciences, Lagos State Polytechnic, Ikorodu, Nigeria

3 Department of Chemical Engineering, Yaba College of Technology, Yaba, Lagos, Nigeria

4 Department of Civil Engineering, Lagos State Polytechnic, Ikorodu, Lagos, Nigeria

5 Department of Mechanical Engineering, Lagos State Polytechnic, Ikorodu, Lagos, Nigeria

*Address all correspondence to: olufeday@gmail.com; fadayini.o@mylaspotech.edu.ng

References

[1] N. A. Madlool, R. Saidur, M. S. Hossain, and N. A. Rahim, "A critical review on energy use and savings in the cement industries," *Renew. Sustain. Energy Rev.*, vol. 15, no. 4, pp. 2042–2060, 2011, Accessed: Nov. 08, 2020. [Online]. Available: https://ideas.repec.org/a/eee/rensus/v15y2011i4p2042-2060.html.

[2] J. P. John, "Parametric Studies of Cement Production Processes," *J. Energy*, vol. 2020, 2020, doi: 10.1155/2020/4289043.

[3] O. S. Ohunakin, O. R. Leramo, O. A. Abidakun, M. K. Odunfa, and O. B. Bafuwa, "Energy and Cost Analysis of Cement Production Using the Wet and Dry Processes in Nigeria," *Energy Power Eng.*, vol. 05, no. 09, pp. 537–550, 2013, doi: 10.4236/epe.2013.59059.

[4] C. Galitsky and L. Price, "Opportunities for Improving Energy Efficiency, Reducing Pollution and Increasing Economic Output in Chinese Cement Kilns," *ACEEE 2007 Summer Study Energy Effic. Ind.*, pp. 1–12, 2007.

[5] P. Oladunjoye, "Nigeria: Cement Production And Survival of the Construction Industry - allAfrica.com," *Daily Independent*, Jun. 04, 2011. https://allafrica.com/stories/201106061192.html (accessed Nov. 08, 2020).

[6] W. T. Choate, "Energy and Emission Reduction Opportunities for the Cement Industry," *Energy Effic. Renew. Energy*, pp. 1–41, 2003.

[7] M. Schneider, M. Romer, M. Tschudin, and H. Bolio, "Sustainable cement production-present and future," *Cement and Concrete Research*, vol. 41, no. 7. Elsevier Ltd, pp. 642–650, 2011, doi: 10.1016/j.cemconres.2011.03.019.

[8] M. G. Rasul, W. Widianto, and B. Mohanty, "Assessment of the thermal performance and energy conservation opportunities of a cement industry in Indonesia," *Appl. Therm. Eng.*, vol. 25, no. 17–18, pp. 2950–2965, 2005, doi: 10.1016/j.applthermaleng.2005.03.003.

[9] P. A. Alsop, *The Cement Plant Operations Handbook*, Third. Portsmouth: David Hargreaves, International Cement Review, 2001.

[10] Portland Cement Association, "Report on Sustainable Manufacturing," 2006. [Online]. Available: www.cement.org/smreport11.

[11] R. G. Bond and C. P. Straub, *CRC handbook of environmental control*, vol. 4. Cleveland: CRC Press, 1974.

[12] WSP Parson Brinkerhoff and DNV GL, "Industrial Decarbonisation & Energy Efficiency Roadmaps to 2050: Cement," 2015.

[13] Engin & V. Ari, Energy auditing and recovery for dry type cement rotary kiln systems. *Energy Conversion and Management*, 46, 2005

[14] D. Gielen, & P. Taylor, Indicators for industrial energy efficiency in India. *Energy*, 34, 2009

[15] C. Sheinbaum, & I. Ozawa, Energy use and CO2 emissions for Mexico's cement industry. *Energy 23(9)*, 725–32, 1998

[16] J. Soares, & M. Tolmasquim, Energy efficiency and reduction of CO2 emissions through 2015. *Mitigation and Adaptation Strategies for Global Change, 5*, 297–318, 2000

[17] E. Worrell, & N. Martin, Potentials for energy efficiency improvement in the US cement industry. *Energy, 25*, 189–214, 2000

[18] IEA (2004), *Coal Information 2004*, OECD Publishing, Paris, https://doi.org/10.1787/coal-2004-en.

[19] Olayinka S. Ohunakin, Oluwafemi R. Leramo, Olatunde A. Abidakun, Moradeyo K. Odunfa, Oluwafemi B. Bafuwa, 2013, "Energy and Cost Analysis of Cement Production Using the Wet and Dry Processes in Nigeria", Energy and Power Engineering, 2013, 5, 537-550

[20] N. A. Madlool, R. Saidura, M. S. Hossaina, and N. A. Madloo Rahim, "A Critical Review on Energy Use and Savings in the Cement Industries," Renewable and energy Reviews, Vol. 15, No. 4, 2011, pp. 2042-20 http://dx.doi.org/10.1016/j.rser.2011.01.005

[21] http://www.nigerianbestforum.com/blog/?p=59984

[22] A. Avami and S. Sattar, "Energy Conservation Opportunities: Cement Industries in Iran," International Journal of Energy, Vol. 3, 2007, pp. 101-110.

[23] globalpetrolprices.com. (2021, January 15). *Nigeria Diesel Prices, liter, 11 - Jan - 2021.* Retrieved from GlobalPetrolPrices.com: https://www.globalpetrolprices.com/Nigeria/diesel_prices/

[24] Market Insider. (2021, January 15). *Natural Gas (Henry Hub)*. Retrieved from Market Insider: https://markets.businessinsider.com/commodities/natural-gas-price

[25] Market Insider. (2021, January 15). Coal. Retrieved from Market Insider: https://markets.businessinsider.com/commodities/coal-price

[26] Worrell, K. Kermeli, & C. Galitsky, *Energy Efficiency Improvement and Cost Saving Opportunities for Cement Making.* Utrecht, United states ENERGY STAR, 2013

Chapter 8

Accelerated Carbonation Curing as a Means of Reducing Carbon Dioxide Emissions

Hilal El-Hassan

Abstract

Globally, carbon dioxide concentration has immensely increased post the industrial revolution. With more greenhouse gases generated from human activities, more radiation is being absorbed by the Earth's atmosphere, causing an increase in global temperature. The phenomenon is referred to as the greenhouse gas effect. Alone, the cement industry contributes to approximately 5–8% of the global greenhouse gas emissions. Scientists and environmentalists have proposed different scenarios to alleviate such emissions. Among these, accelerated carbonation curing has been advocated as a promising mechanism to permanently sequester carbon dioxide. It has been applied to numerous construction applications, including concrete masonry blocks, concrete paving blocks, ceramic bricks, concrete pipes, and cement-bonded particleboards. Experimental results have shown that not only does it significantly reduce the carbon emissions, it also improves the mechanical and durability properties of carbonated products. The process enhances material performance, offers environmental benefits, and provides an excellent means to recycle carbon dioxide.

Keywords: carbonation curing, construction applications, mechanical properties, durability performance, environmental benefits

1. Introduction

Greenhouse gases are responsible for maintaining ecological balance and warmth on the planet. Of the total greenhouse gases, carbon dioxide is the main component comprising about 76% [1, 2]. With more CO_2 generated from industries, urbanization, and human activities, more radiation will be absorbed by the Earth's atmosphere, causing an increase in global temperature. The phenomenon is referred to as the greenhouse gas effect. In the 1990s, the rise in the planet's average temperature was 0.74°C. By the end of the 21st century, it is projected to increase by up to 6.4°C [3], instigating cataclysmic changes, as melting of polar ice, increase in sea levels, variations in rainfall and relative humidity (RH), and disappearance of fauna, among others [4].

Of the emitted carbon dioxide gas, the cement industry is responsible for about 5–8% [5]. Such emission is associated with the calcination of limestone ($CaCO_3$) to produce lime (CaO) and CO_2 and the burning of fossil fuels for clinkering and grinding. Indeed, it is estimated that the production of one ton of cement releases an equal weight of CO_2 gas [6]. Cement is the main constituent of concrete, the world's most consumed man-made material with approximately one cubic meter being produced per capita [7]. With the rise in the human population, there is an ever-increasing demand for infrastructure and superstructures. Accordingly, more cement and concrete will be needed. As a result, cement production is becoming an increasing global pressing issue from an ecological, social, and environmental standpoint. To alleviate the emission of CO_2 associated with producing cement and concrete, scientists and environmentalists have proposed different schemes, including the replacement of cement with supplementary cementitious materials (SCMs), increase in energy efficiency, use of alternate fuels, and carbon sequestration [8].

The first scheme proposes modifications to the mixture proportions by replacing cement with SCMs, which are typically industrial waste materials. Cement kiln dust, a by-product of cement manufacturing, has been utilized in producing sustainable composites for construction applications with economic and environmental benefits [9–11]. Other industrial by-products, including fly ash, slag, rice husk ash, limestone filler, and silica fume, have also been used as partial cement replacement in the production of sustainable concrete [12–17]. The properties of so-produced concrete are equivalent, if not superior, to those of conventional cement-based counterparts. Furthermore, efforts have been made to fully replace cement in mortar and concrete. The resulting product has been denoted as alkali-activated or geopolymer mortar/concrete. Numerous studies have investigated the fresh and hardened properties of this novel material and have provided valuable input on its contribution to sustainable construction [18–43]. Nevertheless, the availability and innate compositional variability of the industrial by-products pose a challenge to the adoption and progression of this CO_2 mitigating strategy.

Cement-related carbon emissions could also be reduced by increasing energy efficiency or utilizing alternative fuels during cement production. The use of blended cements, high-activation grinding, and high-efficiency separators, driers, calciners, and clinker cools have been reported to significantly reduce the CO_2 emissions and energy requirements by the cement industry [44–47]. Yet, the suggested modifications in this scheme may not always be practical or economically feasible. Conversely, some studies aimed to alleviate the carbon emissions associated with the generated thermal energy by replacing fossil fuels with alternative fuels, including scrap tires, biomass residues, waste oils, plastics, slaughterhouse residues, spent pot lining, and sewage sludge [48–53]. The scenario is considered environment-friendly, as it conserves natural resources and recycles industrial wastes [54]. However, the different characteristics of these alternative fuels compared to fossil fuels have led to uneven heat distribution, unstable precalciner operation, and dusty kilns, among other complications [55].

While CO_2 emissions could be substantially reduced using the first three methods, they may not always be practical, feasible, or reliable. On the other hand, carbon sequestration or storage has been shown to be a more viable scheme due to its applicability to stationary point sources over the short term. Geological and ocean storage have been mainly practiced for the past few decades [56]. Nevertheless, mineral sequestration has shown great potential, specifically through the accelerated carbonation of hardened cement and concrete.

Carbonation is a curing mechanism applied to fresh concrete, i.e. within the first 24 hours after casting. It entails an exothermic chemical reaction between CO_2 and calcium-carrying compounds in cement. Its advantages are three-fold: 1) rapid strength gain, 2) enhanced durability performance, and 3) permanent sequestration of carbon dioxide gas [57, 58]. This chapter summarizes the research and experimental findings of collective studies that have utilized accelerated carbonation in construction applications, including concrete masonry blocks, concrete paving blocks, concrete pipes, reinforced concrete beams, cement-bonded particleboards, and ceramic bricks. Other topics are also covered, comprising the fundamentals, processes, characterization techniques, and environmental benefits of carbonation of concrete. This work aims to shed light on the technical and environmental gains of accelerated carbonation and its applicability to different construction applications as a means of reducing cement-related carbon dioxide emissions.

2. Fundamentals of carbonation

2.1 Reaction kinetics

Carbonation is a physicochemical reaction between cement and carbon dioxide gas in aqueous conditions. More specifically, it is the calcium silicates and their hydration products that undergo carbonation. At early age, calcium silicates, in the form of alite ($3CaO{\cdot}SiO_2$, or C_3S) and belite ($2CaO{\cdot}SiO_2$, or C_2S), react with CO_2 in the presence of water to produce calcium silicate hydrate ($xCaO{\cdot}SiO_2{\cdot}yH_2O$, or C-S-H) and calcium carbonate ($CaCO_3$). The reaction is primarily dependent on the rate of CO_2 diffusion, which, in turn, is controlled by the concentration of CO_2 and its pressure during carbonation [59]. The exothermic reactions are shown in Eq. (1) and (2) [60, 61].

$$3(3CaO{\cdot}SiO_2)+(3-x)CO_2+yH_2O \rightarrow xCaO{\cdot}SiO_2{\cdot}yH_2O+(3-x)CaCO_3 \quad (1)$$

$$2(2CaO{\cdot}SiO_2)+(2-x)CO_2+yH_2O \rightarrow xCaO{\cdot}SiO_2{\cdot}yH_2O+(2-x)CaCO_3 \quad (2)$$

Carbonation is an accelerated hydration reaction; yet, it is different than typical hydration of cement with water, as calcium carbonate formed instead of calcium hydroxide ($Ca(OH_2)$) [60, 61]. In Eq. (1) and (2), the values of x and y depend on the extent of reaction, whereby the theoretical maximum degree of reactivity is 50% of the cement mass [62]. In the event of extensive carbonation exposure, CO_2 may decalcify the newly-formed C-S-H to produce silica gel (SiO_2) and calcium carbonate, as per Eq. (3) [63]. However, this reaction is not likely to occur in the short-term accelerated carbonation curing regimes employed in the studies addressed herein.

$$xCaO{\cdot}ySiO_2{\cdot}zH_2O+xCO_2 \rightarrow xCaCO_3+y(SiO_2{\cdot}tH_2O)+(z-yt)H_2O \quad (3)$$

Moreover, the carbonation reaction does not consume all C_3S and C_2S particles, allowing for their subsequent hydration in a post-carbonation moist curing environment [64]. Accordingly, the end-result cementitious matrix is an intermix of C_2S, C_3S, $CaCO_3$, C-S-H, and $Ca(OH)_2$ [65–70].

2.2 Characterization of the reaction products

Carbonation of calcium silicates (C_3S and C_2S) is a form of mineral carbon sequestration that converts carbon dioxide gas into thermodynamically stable calcium carbonate. The calcium carbonate has been detected in three polymorph phases, namely aragonite, vaterite, and calcite [61]. Microstructure analysis showed that the first two polymorphs formed due to carbonation of C-S-H, while the third one was a product of carbonating calcium silicates [71, 72]. Yet, among the three, calcite has been predominantly identified as the main reaction product of accelerated carbonation. In fact, thermogravimetric analysis (TGA) and X-ray diffraction (XRD) have shown that poorly crystalline aragonite and vaterite transformed into the more stable crystalline calcite polymorph during subsequent hydration [66, 73]. This phenomenon is shown in the XRD spectra of **Figure 1**.

While calcium carbonate has been highlighted as the main carbonation reaction product, C-S-H gel has been identified on fewer occasions. Using XRD, C-S-H was qualitatively detected as a slight increase in the baseline between 25 and 35°2θ [66, 73]. This C-S-H was similar in its amorphous morphology to that formed during typical C_3S hydration but different in that it was characterized by a lower

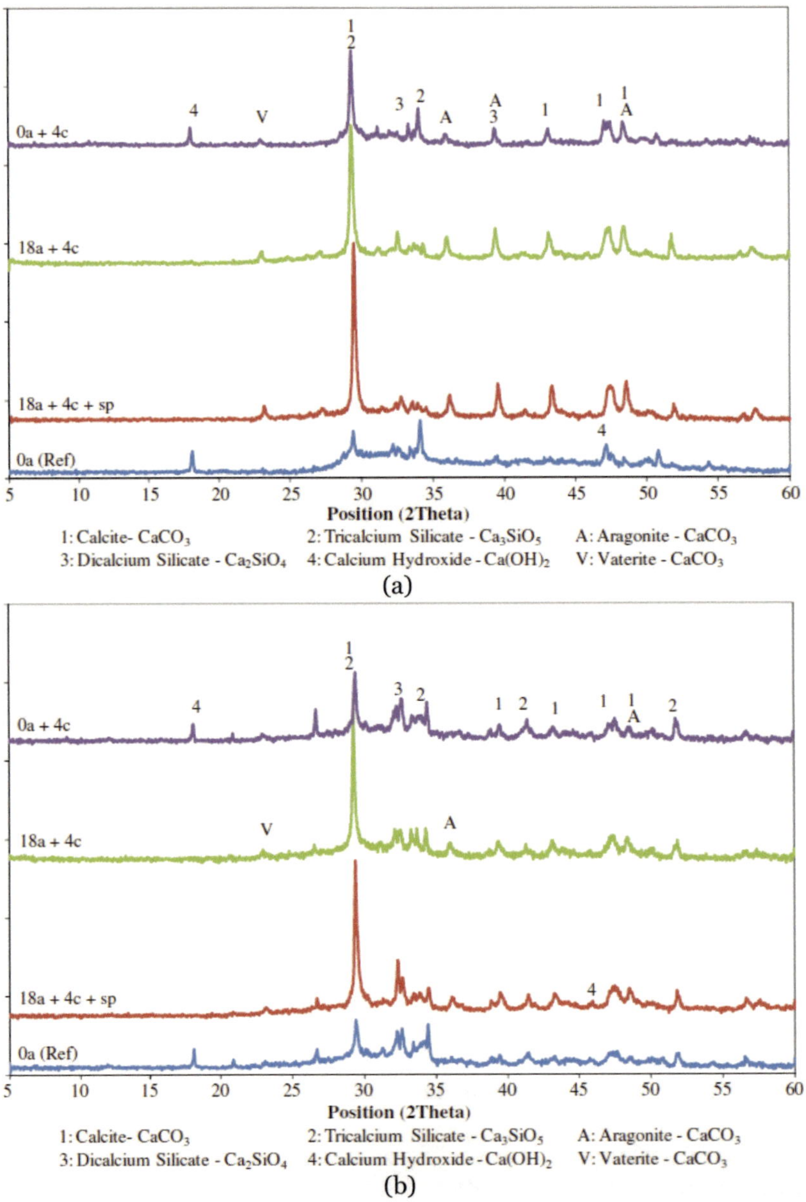

Figure 1.
XRD pattern of concrete hydration- and carbonation-cured concrete at the age of (a) 1 day and (b) 28 days [66]. Reproduced with permission from the publisher.

CaO-to-SiO_2 ratio [71, 74, 75]. Actually, it was believed that a high degree of carbonation reaction (carbon uptake exceeding 18%, by cement mass) led to intermixing amorphous C-S-H with dominant $CaCO_3$ to form a calcium silicate hydrocarbonate product, rendering it difficult to be distinctively identifiable [66, 71]. Conversely, lower reactivity (carbon uptake below 10%, by cement mass) integrated small $CaCO_3$ crystals into a C-S-H-dominant nanostructure [65].

The morphology of these reaction products has also been studied. Calcium carbonate, in its three polymorphs, was reported in different shapes. Cubic, crystal shapes were identified, as depicted in **Figure 2**, when ordinary Portland cement (OPC) concrete was exposed to simultaneous carbonation and chloride ion ingress [76]. Successive preconditioning and carbonation curing of OPC paste and concrete presented amorphous C-S-H gel, calcium hydroxide hexagons, and amorphous calcium carbonate, as shown in **Figure 3a** [65]. A similar morphology is illustrated in **Figure 3b**, whereby OPC concrete made with drinking water treatment sludge was carbonated for 20 hours after 4 hours of preconditioning [77]. Further, an amorphous microstructure with a matrix comprising C-S-H and $CaCO_3$ was reported when OPC concrete was carbonated for 4 hours after 18 hours of preconditioning (**Figure 4a**). Similar findings have been reported in other studies [68, 78, 79]. Conversely, the morphology of Portland limestone cement (PLC) concrete exposed to a similar carbonation scheme encompassed ball-like forms covered with sharp crystals, as presented in **Figure 4b** [80]. Compared to carbonated OPC concrete, the microstructure of counterparts made with PLC was more porous with higher degree of crystallinity. It was believed that the presence of fine limestone in PLC may have served as nucleation sites for calcium carbonate crystal growth [80].

3. Carbonation process

With the advanced understanding of accelerated carbonation, more research has adopted carbonation curing for precast concrete products. The process promises to

Figure 2.
Morphology of cementitious matrix exposed to simultaneous carbonation and chloride ioningress [76]. Reproduced with permission from the publisher.

Figure 3.
Morphology of cementitious matrix exposed to (a) 18-hour preconditioning and 2-hour carbonation [18], (b) 4-hour preconditioning and 18-hour carbonation [30]. Reproduced with permission from the publisher.

alleviate anthropogenic emissions through a mineral carbon sequestration technique. However, the environmental impact of carbonation is related to the degree of reaction, which is a function of the availability of water and pore precipitation sites. As such, different curing regimes have been adopted to optimize the amount of water for the highest reaction efficiency. These curing regimes were somewhat different in the adopted duration, temperature, and relative humidity. Yet, they had commonly implemented a three-phase curing process, namely preconditioning, carbonation curing, and post-carbonation hydration.

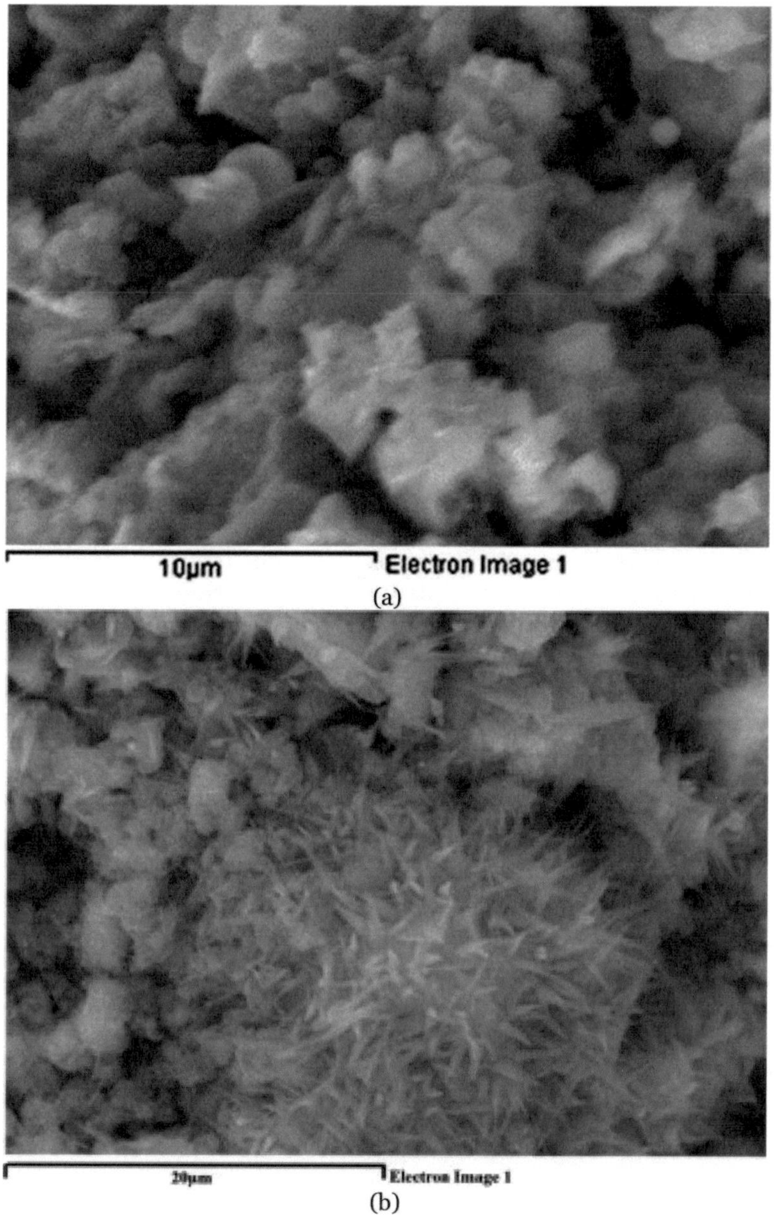

Figure 4.
SEM micrograph of carbonated (a) OPC concrete [66], (b) PLC concrete [80]. Reproduced with permission from the publisher.

3.1 Preconditioning

Past studies have reported that free water was necessary to facilitate the dissolution of CO_2, however excess water obstructed its penetration through the available porous path [66]. As such, preconditioning was introduced to optimize the amount of water prior to exposing the designated samples to carbon dioxide gas. The adoption of such a process led to a superior carbonation degree and enhanced long-term hydration [81].

For dry mixes, preconditioning took place immediately after casting and before the initial setting of the mix [65, 70, 73, 80, 82]. Conversely, wet mixes were only preconditioned after the initial setting [67, 83]. In general, the conditions comprised a duration, temperature, and relative humidity in the ranges of 2–24 hours, 20–25°C, and 40–60%, respectively [66, 67, 70, 73, 80, 82–86]. The effect of preconditioning at 25°C and 50% relative humidity on the water content during 14-day preconditioning is illustrated in **Figure 5**. To simulate industrial practice and limit the total curing time window to 24 hours, a maximum preconditioning duration of 18 hours was recommended [82]. At the end of the preconditioning phase, the cementitious matrix would include anhydrous and hydrated calcium silicate compounds.

3.2 Carbonation curing

Carbonation curing encompasses the time period in which concrete is exposed to carbon dioxide gas. Typically, CO_2 is released into a closed chamber and left for a certain duration and under specific conditions for the reaction to take place. A static carbonation system has been typically adopted by most researchers, as shown in **Figure 6a** [66, 70, 80, 82, 85, 87–93]. This carbonation scheme utilized a closed system, whereby the water that evaporated due to the exothermic reaction was included in the estimation of the degree of carbonation. Nevertheless, the reaction was hindered through the precipitation of calcium carbonate particles in the available porous space, leading to a decrease in porosity and retarded diffusivity. To overcome this challenge, a pseudo-dynamic carbonation setup was devised (**Figure 6b**) [73, 94]. This system removed surface free water in a controlled environment and enhanced carbon dioxide penetration by creating a route of capillaries through the sample. It is worth noting that both systems employed a vacuum prior to injecting 99%-pure CO_2 and the pressure was set to 1 bar. Several other researchers used a flue gas to enhance the environmental impact of carbonation, however, the degree of reaction was lesser [67, 95, 96]. Higher pressures of up to 5 bars were also employed [70, 84, 85, 97–99]. Although some promising results were reported when carbonation was utilized at higher pressure, the applicability and feasibility of adopting pressurized carbonation by the industry are yet to be evaluated.

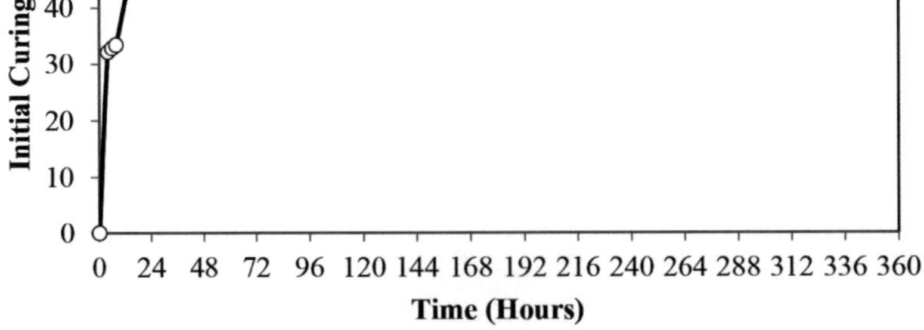

Figure 5.
Water loss during preconditioning of lightweight concrete [82]. Reproduced with permission from the publisher.

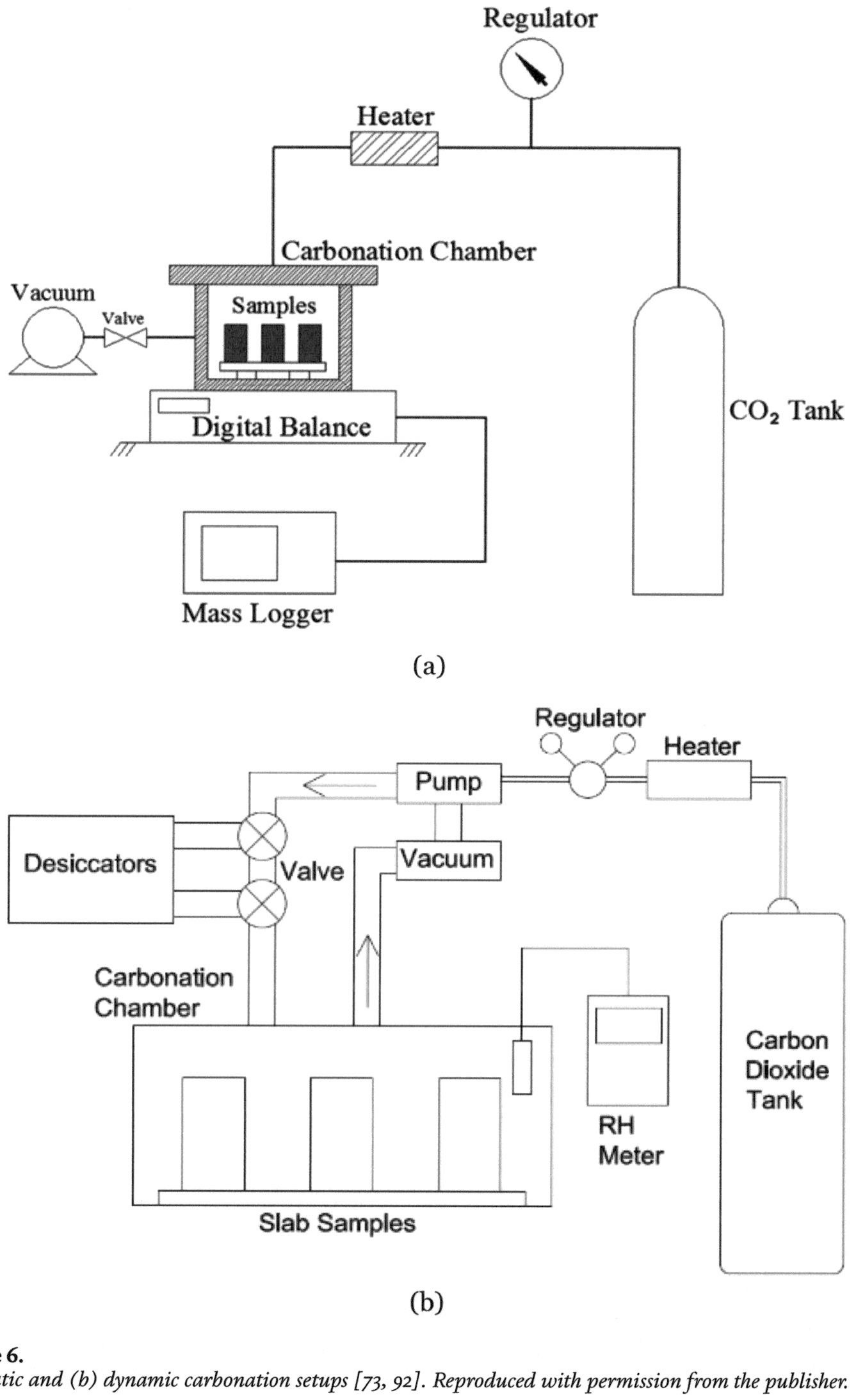

Figure 6.
(a) Static and (b) dynamic carbonation setups [73, 92]. Reproduced with permission from the publisher.

3.3 Post-carbonation hydration

The third phase of the carbonation process is the post-carbonation hydration. This step is critical to restore the water lost during preconditioning and the exothermic carbonation reaction and to promote subsequent hydration of unreacted hydraulic cement phases. Early research has reported up to 45% increase in the compressive strength when samples were placed in water for 3 days after carbonation [64]. Other work incorporated spraying 4-hour carbonated concrete samples every other day until the age of 7 days [82]. The compressive strength increased by 20% compared to carbonated samples left to cure in open air. It is believed that such improvement in mechanical properties is primarily owed to the enhanced pore structure [84].

4. Carbonation degree and characterization techniques

Experimental research findings have provided evidence of the feasibility of utilizing carbonation curing for precast concrete products. Yet, the construction industry has not widely adopted it. To promote its utilization and adoption, most past studies aimed to augment the environmental benefit by maximizing the degree of carbonation reaction, which was typically characterized by the carbon uptake. One way to measure the carbon uptake was by examining the mass gained during the carbonation period, assuming homogeneous carbonation across the sample. Because the system was treated as a closed one, the water lost during the exothermic carbonation reaction was collected and added to the final sample mass. Based on Eq. (4), the carbon uptake is the difference in mass between before and after carbonation with the addition of the mass of water lost as a function of the mass of cement. This method has been implemented in several past studies [60, 66, 70, 73, 74, 80, 82, 89, 100–102].

$$\text{Carbon uptake}(\%) = \frac{\text{Final mass} - \text{Initial mass} + \text{water lost}}{\text{Mass of cement}} \text{x}\,100\% \tag{4}$$

Thermal analysis is another means of measuring the absolute carbon uptake. In this technique, a thermogravimetric analyzer (TGA) was utilized to monitor the mass loss of a powder sample of carbonated concrete with heat [103]. Alternatively, concrete chunks were decomposed in an electrical muffle furnace by raising the temperature from 25°C to approx. 1000°C. Within this range, multiple hydration and carbonation products were decomposed. The temperature ranges 105–200°C, 200–420°C, 420–550°C, 550–720°C, and 720–950°C were associated with the decomposition of low-temperature C-S-H and ettringite, well-formed C-S-H and C-A-H, calcium hydroxide, poorly crystalline calcium carbonate (vaterite and aragonite), and well crystalline calcite, respectively [66, 71, 73, 80, 81, 104, 105]. However, these ranges slightly differed depending on the type of binder used and the mixture proportions in general, and may even result in an overlap between carbonates and hydrates. To overcome this problem, Fourier transform infrared spectroscopy (FTIR) was employed alongside TGA, as shown in **Figure 7** [106]. It is a vibrational spectroscopic analytical technique that could detect calcium carbonate from the C-O characteristic peak at a wavelength of 1415 cm^{-1} [106]. Other analytical tools have also been utilized together with TGA to identify carbonation products, including nuclear magnetic resonance (^{29}Si NMR) and XRD [65].

Carbon uptake was also determined by employing coulometric titration in a solution of hydrochloric acid [107]. The technique involved submerging a carbonated concrete powder in the acid solution and measuring the released carbon using the coulometer. The carbon uptake was then obtained using stoichiometric proportions. It is worth noting that thermal analysis and coulometric titration decompose all the carbonates present in the concrete. In an attempt to improve the sustainability of cement, some manufacturers have been replacing

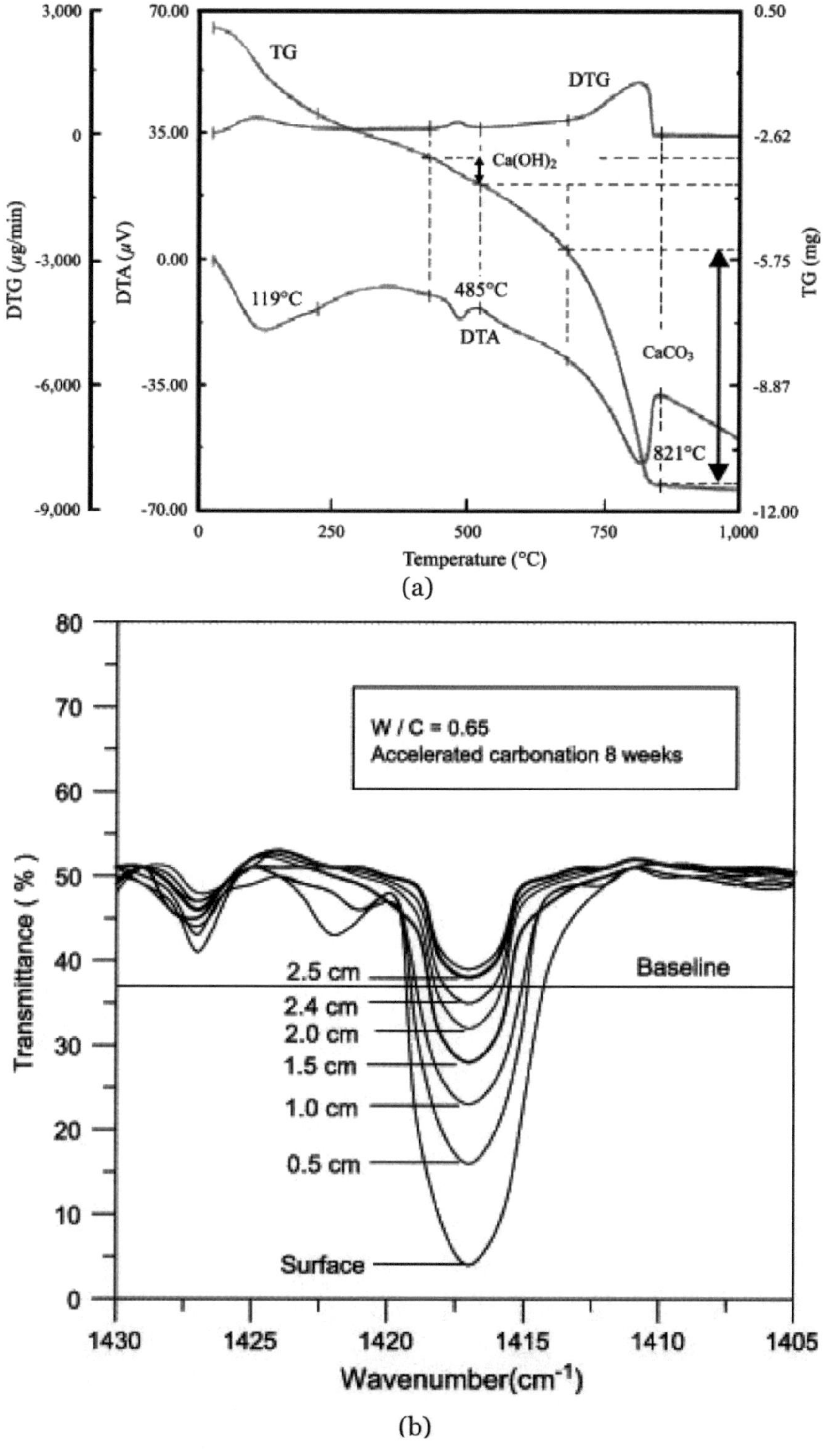

Figure 7.
(a) Thermogravimetric curves and (b) FTIR spectra of carbonated concrete [106]. Reproduced with permission from the publisher.

cement with certain amounts of limestone powder. Such limestone should be deducted from the overall measured carbon content to obtain the absolute carbon uptake [80, 108].

5. Carbonation in construction applications

Carbonation curing has been investigated for precast concrete products as a sustainable alternative curing regime to the more typically used steam and moist curing techniques. This section summarizes the collective studies that have examined the effect of accelerated carbonation on the performance of concrete masonry blocks, concrete paving blocks, concrete pipes, reinforced concrete beams, cement-bonded particleboards, and ceramic materials.

5.1 Concrete masonry blocks

Accelerated carbonation has been examined as a sustainable curing technique to replace steam curing for concrete blocks. Past studies have examined the mechanical and durability properties of concrete masonry units made with OPC and cured following 0- to 18-hour initial air curing and 2- to 4-hour static carbonation [82, 86, 87, 101, 109, 110]. The carbon uptake reached up to 24%, by cement mass, representing a 48% degree of reaction. The compressive strength within 1 day and at 28 days was comparable to that of steam- and moist-cured counterparts, with values reaching up to 10 and 39 MPa, respectively. It is worth noting that the highest carbon uptake and strength results were noted for samples that were preconditioned for 18 hours in open air prior to carbonation [82, 86, 109, 110]. Additionally, carbonation improved the resistance to chloride penetration by 1.4 and 6.2 times compared to the two conventionally-cured concrete, respectively, and enhanced resistance to sulfate attack by at least 1.5 times [101]. Also, carbonation-cured concrete (18a + 4c and 18a + 4c + sp) had 2 to 3 times better freeze-thaw resistance than concrete cured using steam (2a + 4 s) or moist curing (0a), as shown in **Figure 8** [87].

When OPC was replaced by PLC, similar trends related to mechanical properties were noted, but the strength was ultimately lower due to a more porous and crystalline microstructure, as noted in **Figure 4** [66, 80]. Furthermore, concrete blocks

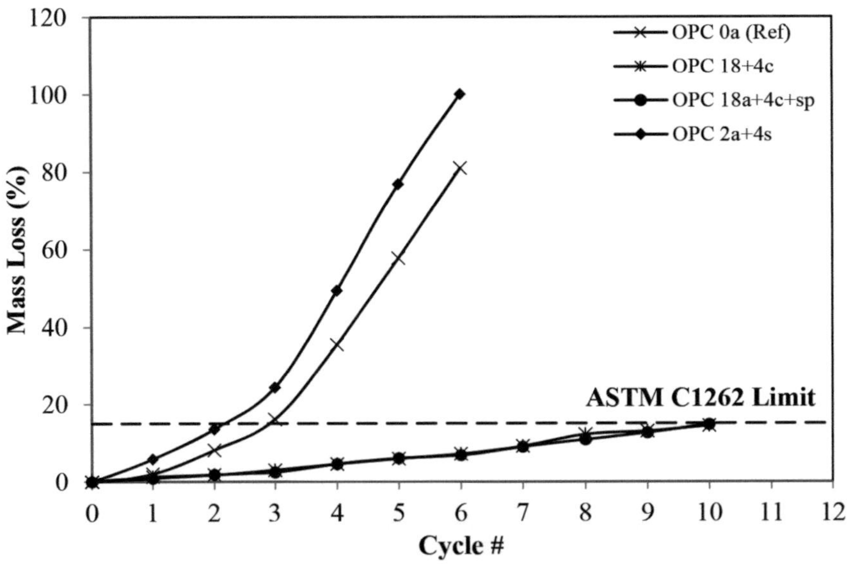

Figure 8.
Freeze-thaw resistance of carbonated and hydrated concrete [87]. Reproduced with permission from the publisher.

were made with partial replacement of fine aggregates with drinking water treatment sludge and subjected to carbonation [77, 111]. The 1- and 28-day compressive strengths of carbonation-cured concrete blocks were up to 273 and 42% higher those that of normally-cured counterparts, respectively. Splitting tensile strength of the former was also higher than the latter but by no more than 45%. The durability of the concrete blocks was improved, evidenced by the reduction in water capillary absorption and better resistance to sulfate attack. It is believed that this enhancement in durability performance was owed to the pore-filling capacity of newly-formed calcium carbonate [77, 111]. The only detrimental effect of carbonation curing was the increased leaching of aluminum and copper ions, especially for the first 3 days. Nevertheless, the total 60-day leaching concentrations were within the acceptable range, indicating that carbonated concrete blocks made with drinking water treatment sludge were environment-friendly construction materials [111].

5.2 Concrete paving blocks

Paving blocks are precast non-reinforced concrete products used in construction applications, including pedestrian and vehicle pavements. With no steel reinforcement and the ability to mass-produce in a precast concrete plant, it is an ideal construction product that could sequester CO_2 through accelerated carbonation curing. In one study, Wang, Yeung [102] examined the use of CO_2 curing to create high performance, low-carbon paving blocks made with contaminated sediment and binary cement. Concrete samples were left to cure in a waterproof membrane until test age, after which they were placed in a drying chamber for 4 hours and cured with CO_2 gas for 24 hours at 0.1 bar above atmospheric pressure. Results of **Figure 9** show that the compressive strength of carbonation-cured concrete blocks was at least 2 times higher than that of air-cured counterparts. Evidently, carbonation curing accelerated the transformation of anhydrous phases into carbonates, while also promoting the formation of more hydrates during subsequent hydration.

Accelerated carbonation has also been employed to cure concrete paver blocks using pure CO_2 and flue gas [112–114]. After preconditioning, concrete samples were exposed to 2 to 4-hour carbonation and then, placed in a mist room to promote subsequent hydration up to 28 days. The CO_2 uptake was reported to be 3.29

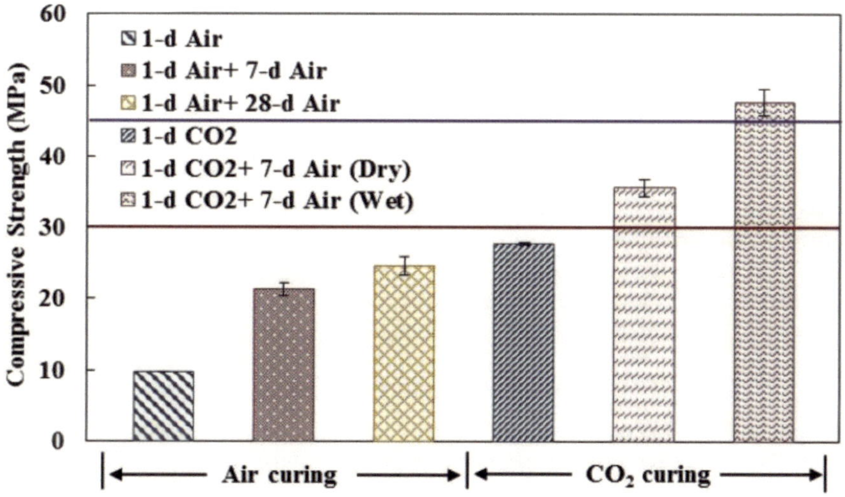

Figure 9.
Compressive strength of concrete paving blocks with various curing methods [102]. Reproduced with permission from the publisher.

and 10.38%, by cement mass, for samples that were carbonated in 20 and 99% CO_2, respectively. Such lower uptake in the former was due to the lower CO_2 concentration, leading to less $CaCO_3$ formation and lower compressive strength than the latter. This also resulted in higher water absorption and inferior resistance to efflorescence [112, 113]. In addition, Shao and Lin [114] reported up to 60 times more freeze–thaw resistance when concrete paver blocks were carbonated rather than hydrated.

5.3 Concrete pipes

Past research has shown that carbonation curing is best applied to fresh concrete directly after casting to promote the chemical reaction between calcium silicates and CO_2 gas. Wet mixes are problematic when demolding within the first few minutes, while dry mixes are ideal for such applications. Concrete pipes are among the different types of concrete products that utilize dry mixes with zero slump. Accordingly, carbonation curing of concrete pipes has been investigated [114]. Samples were cast with a water-cement ratio (w/c) of 0.26 with cement, coarse aggregate, and fine aggregate contents of 426, 853, and 853 kg/m^3. Directly after casting, they were demolded and placed in a carbonation chamber for 2 hours at a pressure of 1.5 bar and CO_2 purity of 99%. The average carbon uptake was found to be 11.3%, by cement mass. Compared to the hydrated control samples, the carbonated counterparts had a similar 28-day compressive strength of 16 MPa. As concrete pipes may be reinforced with steel, the pH of the carbonated concrete was measured. It was interesting to note that the pH remained above 12, indicating the ability to employ carbonation curing for concrete pipes even in the presence of steel reinforcement.

Other work investigated the feasibility of curing concrete pipes in combined steam and carbonation regime in an attempt to reduce the energy footprint of steam curing, while also sequestering CO_2 [70]. Based on the early-age strength results, concrete pipe samples cured in a combination of the two curing regimes, i.e. steam and carbonation, provided equivalent and superior results to those that were steam and carbonation-cured, respectively, and had a CO_2 uptake of approx. 9%, by cement mass. Compared to samples exposed to steam, those that underwent combined curing showed higher resistance to chloride penetration, sulfate attack, and acid attack, possibly due to the consumption of hydroxyl ions and the formation of calcium carbonate.

5.4 Reinforced concrete beams

Despite its adverse effect on reinforced concrete, carbonation of precast reinforced concrete products may be beneficial if performed at an early age. An early-age carbonation curing process was developed for precast reinforced concrete [90, 115]. The detailed curing regime encompassed i) 5-hour in-mold curing at 25°C and 60% RH, ii) 5–6-hour off-mold preconditioning at 25°C and 50 ± 5% RH, iii) 12-hour carbonation curing at a pressure of 5 bars, and iv) 27-day subsequent hydration at 25°C and 95% RH. Carbon sequestration potential was characterized by the CO_2 uptake. It increased from 8 to 15% as the pressure increased from 1 to 5 bar, respectively. This was also associated with an increase in carbonation depth from 8 to 17 mm. Although carbonation decreased the pH of the surface at early age to 9.2, it could recover to 12.3 after 27-day subsequent hydration, evident by the phenolphthalein color profile of **Figure 10**. Evidently, the pH of the area surrounding the steel reinforcement was not affected by carbonation. This indicated that the suggested carbonation curing process posed no risk of corrosion to the

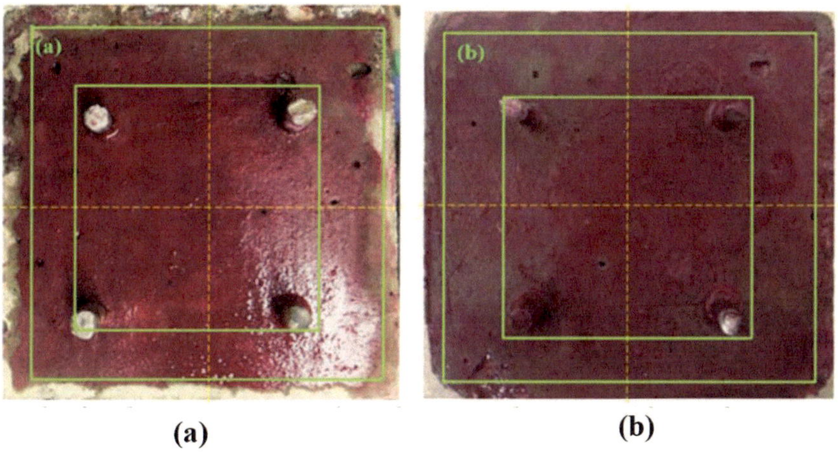

Figure 10.
Depth of carbonation in a reinforced concrete beam: (a) after carbonation curing; (b) after carbonation curing and 28-day subsequent hydration [92]. Reproduced with permission from the publisher.

steel reinforcement and could be adopted for precast reinforced concrete products. Furthermore, carbonation-cured samples exhibited higher compressive strength, surface resistivity, resistance to chloride penetration, and resistance to weathering carbonation than hydration-cured counterparts. Apparently, carbonation curing reduced the pore size and volume due to calcium carbonate formation and precipitation within the cementitious matrix.

5.5 Cement-bonded particleboards

Cement-bonded particleboards are construction products that incorporate cement and fine wood chip fractions. As cement is the main binder in such products, past studies have investigated the use of carbonation curing as a replacement to typical hydration to expedite the production process, while also providing a sink for carbon sequestration. Early studies showed that 2-hour carbonation resulted in a carbon uptake of up to 24%, by cement mass, and compressive strength of 10.5 MPa, which was three times that of the hydrated reference [113, 116]. Nevertheless, prolonging carbonation to 24 hours enhanced the reaction efficiency to obtain an uptake of up to 28% [117]. The resulting flexural strength, freeze-thaw resistance, and wet-dry durability were higher than conventionally-cured counterparts.

Another study employed a wetting-drying-carbonation curing scheme for cellulose fiber-reinforced cement boards [69]. Experimental results showed that accelerated carbonation curing was beneficial to the performance of cement boards. Compared to conventional water curing, it provided superior flexural strength and toughness and reduced capillary porosity and microcracking in the autoclave. Similar findings were reported when cement-bonded particleboards were subjected to supercritical CO_2 curing [118, 119].

In addition to the utilization of carbonation curing to sequester carbon dioxide in cement-bonded particleboards, it has been employed to promote the recyclability of waste materials [120–123]. The mechanical, durability and physical properties of carbonated cement-bonded particleboards were comparable, if not superior, to those of air-cured and hydrated counterparts. Such performance enhancement is owed to the ability of carbonation curing to improve the intactness at the cement-fiber interface, limit the interfacial microcracks, and occupy the capillary space with newly-formed calcium carbonate.

5.6 Ceramic bricks

The applicability of accelerated carbonation curing has been explored in numerous construction applications. The common factor among these applications is the carbonation of calcium or magnesium silicates to produce carbonates. Ceramic materials are rich in such silicates and may be carbonated upon exposure to CO_2. As such, accelerated carbonation was applied to ceramic bricks from Andalusian factories in Spain [124]. The curing process entailed 24- to 720-hour exposure to CO_2 at a pressure of 10 bars. The authors noted that longer exposure led to higher carbon uptake, with values reaching up to 10%, by ceramic weight. These results highlight the possibility of employing carbonation curing to ceramic waste materials as a means of permanently sequestering carbon dioxide. Yet, more research is needed to validate the findings and evaluate the feasibility of adopting such a technique by the industry.

6. Environmental benefit

Concrete construction applications serve as a potential carbon dioxide sink for CO_2 sequestration. Rather than disposing of CO_2 in geological sites, it can be recycled into concrete with the added benefit of early-age strength and improved durability performance. Concrete products that are typically cured using the steam curing regime can be carbonated to relieve the dependency on high pressure and temperature steam. For instance, a concrete block can sequester nearly 0.5 kg of CO_2, at an uptake of 24%, by cement mass. At a global annual production of 1800 billion concrete blocks and bricks [125], it will be possible to sequester 900 million tons of CO_2, which is equivalent to carbon sequestration in approx. 900 geological sites. In comparison, a single concrete paver block could sequester 15.3 g of CO_2, characterized by an update of 10.4%, by cement mass. With 51.4 billion concrete paver blocks (assuming 20% cement content, a thickness of 80 mm, and a density of 2200 kg/m^3) produced annually [126], these products could sequester up to 1.07 million tons of CO_2.

Concrete pipes are produced on the scale of 62 million tons per year [127]. At a carbon uptake of 20%, the concrete pipe industry can sequester up to 1.2 million tons of CO_2 per year. Further, precast concrete products in the form of railway ties can store a total of 0.1 million tons of CO_2 per year globally [128]. Conversely, the 9.5 billion m^2 of cement-bonded boards produced annually could sequester 10.8 million tons of CO_2, assuming 50% cement content, a thickness of 8 mm, a density of 1500 kg/m^3, and CO_2 uptake of 19%, by cement mass [129]. Although ceramic tiles are different than cementitious concrete, they have also presented a 10% carbon uptake, by total weight. With the global production of 13.6 billion m^2, and assuming a typical thickness of 1.5 cm, the ceramic tile industry would be capable of sequestering 20.4 million tons of CO_2 [130]. Yet, it should be noted that only one study has been conducted in this research area, signifying the need for further investigation. On a global scale, if all producers of the concrete products presented herein were to adopt carbonation curing, a total of 934 million tons of CO_2 could be sequestered. With an annual global cement production of 4.2 billion tons [131], accelerated carbonation curing could reduce the carbon emissions associated with the cement industry by 22.2%. This capacity could further increase if carbonation were adopted for curing various precast reinforced concrete products.

While the environmental benefit in terms of CO_2 sequestration has been addressed in various research studies, the curing-related water consumption in carbonation curing compared to steam and moist counterparts has not been investigated yet. Based on the work of El-Hassan, Shao [82], the only water required in accelerated carbonation curing is that for spraying the sample after carbonation. This water promoted subsequent hydration by compensating for the water lost due to preconditioning and the carbonation reaction. As such, the total water consumed in this curing regime was about 0.085 m^3 per m^3 of concrete. In contrast, moist and steam curing are estimated to consume about 3 and 1 m^3 of water for the same volume of concrete, respectively [132]. Clearly, carbonation curing could be deemed more advantageous than moist and steam curing from a water preservation standpoint.

7. Conclusions

Accelerated carbonation is an innovative curing regime that promises to expedite strength gain, improve durability performance, and permanently sequester CO_2 gas in concrete products. Thus, it has the potential to enhance the sustainability of the construction industry.

Reaction kinetics, processes, and final products are comprehensively reviewed. The main chemical reactions occur between calcium silicates (C_3S and C_2S) in the cement and CO_2 gas to produce calcium silicate hydrate (C-S-H) gel and calcium carbonate ($CaCO_3$). Calcium carbonate was detected in its three polymorphic phases, aragonite, vaterite, and calcite, with the former two and latter being associated with the carbonation of C-S-H and calcium silicates, respectively. Their morphology was typical of amorphous, except for the case of carbonating PLC concrete, whereby sharp highly crystalline crystals formed. Conversely, C-S-H was not as easy to detect. In fact, it was intermixed with calcium carbonates to form an amorphous calcium silicate hydrocarbonate product.

The carbonation process was divided into three main stages, preconditioning, carbonation, and subsequent hydration. The utilization of preconditioning was found essential to optimize the water content and promote a higher degree of carbonation reaction. The optimum relative humidity employed in preconditioning was reported as 50–60%. As for carbonation curing, higher reactivity was noted when higher concentration and pressure of CO_2 were used, evidenced by the higher carbon uptake. Subsequent hydration was introduced afterward to enhance the late age mechanical and durability performance.

The applicability of accelerated carbonation to different construction applications has also been highlighted. Carbonated concrete masonry blocks showed comparable mechanical properties to those of steam- and moist-cured counterparts. Yet, the former's resistance to freeze-thaw damage and sulfate attack was greater than that of the latter. Furthermore, carbonation was applied to concrete paving blocks. The compressive strength and freeze-thaw resistance of carbonated samples were superior to those of hydration- and air-cured equivalents. Similarly, the mechanical and durability performance of concrete pipes and beams subjected to carbonation curing were superior to conventionally-cured counterparts. Also, it was interesting to note that there was no risk of corrosion to the steel reinforcement, as the pH of the surrounding 28-day concrete was above 12. Moreover, the feasibility of employing carbonation as a curing regime for cement-bonded particleboards was assessed. Carbonation curing improved the overall interfacial structure between the cement and fiber and led to the filling of capillary space with newly-formed $CaCO_3$. As a result, enhanced physical, mechanical, and durability properties were reported for carbonated samples compared to conventionally-cured samples. Lastly, carbonation was applied to ceramic bricks as a means of permanently sequestering carbon dioxide.

In addition to its evident improvement in the performance of construction applications, carbonation curing provides a carbon sink to beneficially recycle CO_2. Yet, its full potential can only be attained if it is adopted on a global scale. The application of carbonation curing to all globally-produced concrete blocks, concrete paving blocks, concrete pipes, cement-bonded particleboards, and ceramic bricks can store up to 934 million tons of CO_2, leading to a 22.2% reduction in cement-related carbon emissions. Additionally, it has the potential to reduce the water consumed in moist and steam curing by 97 and 91%, respectively. Evidently, the carbonation curing process enhances material performance, offers environmental benefits, and provides an excellent means to recycle carbon dioxide emitted by the cement industry.

Acknowledgements

This work was financially supported by the United Arab Emirates under grant number 31 N398.

Conflict of interest

The authors declare no conflict of interest.

Author details

Hilal El-Hassan
Civil and Environmental Engineering Department, United Arab Emirates University, Al Ain, United Arab Emirates

*Address all correspondence to: helhassan@uaeu.ac.ae

References

[1] Global greenhouse gas emissions data. https://www.epa.gov/ghgemissions/globalgreenhouse-gas-emissions-data [cited 2020].

[2] Climate change indicators: greenhouse gases. https://www.epa.gov/climateindicators/greenhouse-gases#ref3 [cited 2020].

[3] IPCC. Projections of Future Changes in Climate. Intergovernmental Panel on Climate Change, 2007.

[4] What is the greenhouse effect? http://www.livescience.com/37743-greenhouseeffect.html [cited 2020].

[5] Pacheco-Torgal F, Abdollahnejad Z, Camões AF, Jamshidi M, Ding Y. Durability of alkali-activated binders: A clear advantage over Portland cement or an unproven issue? Constr Build Mater. 2012;30:400-5.

[6] Chindaprasirt P, Chareerat T, Sirivivatnanon V. Workability and strength of coarse high calcium fly ash geopolymer. Cem Concr Compos. 2007;29(3):224-9.

[7] US Geological Survey. Minerals commodity summary - cement USA: USGS; 2016 [cited 2016 October 6].

[8] An update on CO2 capture from cement production. http://www.globalccsinstitute.com/insights/authors/dennisvanpuyvelde/2013/02/20/update-co2-capturecement-production [cited 2020].

[9] Saleh HM, El-Saied FA, Salaheldin TA, Hezo AA. Macro- and nanomaterials for improvement of mechanical and physical properties of cement kiln dust-based composite materials. J Clean Prod. 2018;204:532-41.

[10] Saleh HM, El-Saied FA, Salaheldin TA, Hezo AA. Influence of severe climatic variability on the structural, mechanical and chemical stability of cement kiln dust-slag-nanosilica composite used for radwaste solidification. Constr Build Mater. 2019;218:556-67.

[11] Saleh HM, Salman AA, Faheim AA, El-Sayed AM. Sustainable composite of improved lightweight concrete from cement kiln dust with grated poly(styrene). J Clean Prod. 2020;277:123491.

[12] Samad S, Shah A. Role of binary cement including Supplementary Cementitious Material (SCM), in production of environmentally sustainable concrete: A critical review. International Journal of Sustainable Built Environment. 2017;6(2):663-74.

[13] Panesar DK, Zhang R. Performance comparison of cement replacing materials in concrete: Limestone fillers and supplementary cementing materials – A review. Constr Build Mater. 2020;251:118866.

[14] Kayali O, Sharfuddin Ahmed M. Assessment of high volume replacement fly ash concrete – Concept of performance index. Constr Build Mater. 2013;39:71-6.

[15] Hooton RD. Canadian use of ground granulated blast-furnace slag as a supplementary cementing material for enhanced performance of concrete. Canadian Journal of Civil Engineering. 2000;27(4):754-60.

[16] Khatib JM, Hibbert JJ. Selected engineering properties of concrete incorporating slag and metakaolin. Constr Build Mater. 2005;19(6):460-72.

[17] Juenger MCG, Siddique R. Recent advances in understanding the role of supplementary cementitious

materials in concrete. Cem Concr Res. 2015;78:71-80.

[18] Elahi MMA, Hossain MM, Karim MR, Zain MFM, Shearer C. A review on alkali-activated binders: Materials composition and fresh properties of concrete. Constr Build Mater. 2020;260:119788.

[19] Nawaz M, Heitor A, Sivakumar M. Geopolymers in construction - recent developments. Constr Build Mater. 2020;260:120472.

[20] Amran YHM, Alyousef R, Alabduljabbar H, El-Zeadani M. Clean production and properties of geopolymer concrete; A review. J Clean Prod. 2020;251:119679.

[21] El-Hassan H, Ismail N. Effect of process parameters on the performance of fly ash/GGBS blended geopolymer composites. J Sustain Cem Mater. 2018;7(2):122-40.

[22] Zhang P, Gao Z, Wang J, Guo J, Hu S, Ling Y. Properties of fresh and hardened fly ash/slag based geopolymer concrete: A review. J Clean Prod. 2020;270:122389.

[23] Adesanya E, Ohenoja K, Kinnunen P, Illikainen M. Alkali Activation of Ladle Slag from Steel-Making Process. Journal of Sustainable Metallurgy. 2016;3:300-10.

[24] El-Hassan H, Elkholy S. Performance Evaluation and Microstructure Characterization of Steel Fiber-Reinforced Alkali-Activated Slag Concrete Incorporating Fly Ash. J Mater Civ Eng. 2019;31(10):04019223.

[25] Al-Majidi MH, Lampropoulos A, Cundy A, Meikle S. Development of geopolymer mortar under ambient temperature for in situ applications. Constr Build Mater. 2016;120:198-211.

[26] Aydın S, Baradan B. Mechanical and microstructural properties of heat cured alkali-activated slag mortars. Mater Des. 2012;35:374-83.

[27] Bernal S, De Gutierrez R, Delvasto S, Rodriguez E. Performance of an alkali-activated slag concrete reinforced with steel fibers. Constr Build Mater. 2010;24(2):208-14.

[28] Bernal SA, Rodríguez ED, Mejia de Gutiérrez R, Provis JL, Delvasto S. Activation of Metakaolin/Slag Blends Using Alkaline Solutions Based on Chemically Modified Silica Fume and Rice Husk Ash. Waste Biomass Valorization. 2012;3(1):99-108.

[29] Ismail N, El-Hassan H. Development and Characterization of Fly Ash/Slag-Blended Geopolymer Mortar and Lightweight Concrete. J Mater Civ Eng. 2018;30(4).

[30] Choi S-J, Choi J-I, Song J-K, Lee BY. Rheological and mechanical properties of fiber-reinforced alkali-activated composite. Constr Build Mater. 2015;96(Supplement C):112-8.

[31] Gu YM, Fang YH, Gong YF, Yan YR, Zhu CH. Effect of curing temperature on setting time, strength development and microstructure of alkali activated slag cement. Materials Research Innovations. 2014;18(sup2):S2-829-S2-32.

[32] Gu Y-m, Fang Y-h, You D, Gong Y-f, Zhu C-h. Properties and microstructure of alkali-activated slag cement cured at below- and about-normal temperature. Constr Build Mater. 2015;79:1-8.

[33] Islam A, Alengaram UJ, Jumaat MZ, Bashar II. The development of compressive strength of ground granulated blast furnace slag-palm oil fuel ash-fly ash based geopolymer mortar. Mater Des. 2014;56:833-41.

[34] Islam A, Alengaram UJ, Jumaat MZ, Bashar II, Kabir SMA. Engineering properties and carbon footprint of

ground granulated blast-furnace slag-palm oil fuel ash-based structural geopolymer concrete. Constr Build Mater. 2015;101, Part 1:503-21.

[35] El-Hassan H, Shehab E, Al-Sallamin A. Influence of Different Curing Regimes on the Performance and Microstructure of Alkali-Activated Slag Concrete. J Mater Civ Eng. 2018;30(9):04018230.

[36] Karim MR, Hossain MM, Manjur A Elahi M, Mohd Zain MF. Effects of source materials, fineness and curing methods on the strength development of alkali-activated binder. J Build Eng. 2020;29:101147.

[37] Lee NK, Lee HK. Setting and mechanical properties of alkali-activated fly ash/slag concrete manufactured at room temperature. Constr Build Mater. 2013;47:1201-9.

[38] El-Hassan H, Ismail N, Al Hinaii S, Alshehhi A, Al Ashkar N. Effect of GGBS and curing temperature on microstructure characteristics of lightweight geopolymer concrete. MATEC Web Conf. 2017;120:03004.

[39] Li J, Liu S. Influence of Slag as Additive on Compressive Strength of Fly Ash-Based Geopolymer. J Mater Civ Eng. 2007;19(6).

[40] Ismail N, Mansour M, El-Hassan H. Development of a low-cost cement free polymer concrete using industrial by-products and dune sand. MATEC Web Conf. 2017;120:03005.

[41] Puertas F, Martiez-Ramirez S, Alonso S, Vazquez T. Alkali-activated fly ash/slag cement strength behaviour and hydration products. Cem Concr Res. 2000;30:1625-32.

[42] Provis JL. Alkali-activated materials. Cem Concr Res. 2018;114:40-8.

[43] Elkholy S, El-Hassan H, editors. Mechanical and micro-structure characterization of steel fiber-reinforced geopolymer concrete. Interdependence between Structural Engineering and Construction Management; 2019; Chicago, IL: ISEC Press.

[44] Nidheesh PV, Kumar MS. An overview of environmental sustainability in cement and steel production. J Clean Prod. 2019;231:856-71.

[45] Hasanbeigi A, Price L, Lin E. Emerging energy-efficiency and CO2 emission-reduction technologies for cement and concrete production: A technical review. Renewable and Sustainable Energy Reviews. 2012;16(8):6220-38.

[46] Scrivener K, Martirena F, Bishnoi S, Maity S. Calcined clay limestone cements (LC3). Cem Concr Res. 2018;114:49-56.

[47] Scrivener KL, John VM, Gartner EM. Eco-efficient cements: Potential economically viable solutions for a low-CO2 cement-based materials industry. Cem Concr Res. 2018;114:2-26.

[48] Rahman A, Rasul MG, Khan MMK, Sharma S. Recent development on the uses of alternative fuels in cement manufacturing process. Fuel. 2015;145:84-99.

[49] Kääntee U, Zevenhoven R, Backman R, Hupa M. Cement manufacturing using alternative fuels and the advantages of process modelling. Fuel Processing Technology. 2004;85(4):293-301.

[50] Ghenai C, Inayat A, Shanableh A, Al-Sarairah E, Janajreh I. Combustion and emissions analysis of Spent Pot lining (SPL) as alternative fuel in cement industry. Sci Total Environ. 2019;684:519-26.

[51] Schneider M, Romer M, Tschudin M, Bolio H. Sustainable

cement production—present and future. Cem Concr Res. 2011;41(7):642-50.

[52] Madlool NA, Saidur R, Hossain MS, Rahim NA. A critical review on energy use and savings in the cement industries. Renewable and Sustainable Energy Reviews. 2011;15(4):2042-60.

[53] Puertas F, Blanco-Varela MT. Use of alternative fuels in cement manufacture. Effect on clinker and cement characteristics and properties. Materiales de Construcción. 2004;54(274):51-64.

[54] Trezza MA, Scian AN. Burning wastes as an industrial resource: Their effect on Portland cement clinker. Cem Concr Res. 2000;30(1):137-44.

[55] Chinyama M. P. M. Alternative fuels in cement manufacturing. In: Manzanera M, editor. Alternative Fuel. Rijeka, Croatia: InTech; 2011.

[56] Gluyas J, Mathias S. Geological storage of carbon dioxide book. 1st Edition ed: Woodhead Publishing; 2014.

[57] Young JF, Berger RL, Breese J. Accelerated Curing Of Compacted Calcium Silicate Mortars On Exposure To CO2. Journal of the American Ceramic Society. 1974;57(9):394-297.

[58] Toennies HT, Shideler JJ. Plant Drying Carbonation of Concrete Block - NCMAPCA Cooperative Program. Journal of the American Concrete Institute. 1963;60(5):617-33.

[59] Shi C, Liu M, He P, Ou Z. Factors affecting kinetics of CO2 curing of concrete. J Sustain Cem Mater. 2012;1(1-2):24-33.

[60] Berger RL, Young, J. F., and Leung, K. Accelerated curing of cementitious systems by carbon dioxide - Hydraulic calcium silicates and aluminates - Part II. Cem Concr Res. 1972;2:647-52.

[61] Young JF, Berger, R. L., Breese, J. Accelerated Curing of Compacted Calcium Silicate Mortars on Exposure to CO2. Journal of the American Ceramic Society. 1974;57(9):394-7.

[62] Steinour H. Some Effects of Carbon Dioxide on Mortar and Concrete. American Concrete Institute Journal. 1956;30:905-7.

[63] Papadakis VG, Vayenas, C. G., and Fardis, M. N. Fundamental Modeling and Experimental Investigation of Concrete Carbonation. American Concrete Institute Materials Journal. 1991;88(4):363-73.

[64] Klemm WA, and Berger, R. L. Accelerated curing of cementitious systems by carbon dioxide - Part I - Portland Cement. Cem Concr Res. 1972;2:567-76.

[65] Rostami V, Shao Y, Boyd AJ, He Z. Microstructure of cement paste subject to early carbonation curing. Cem Concr Res. 2012;42(1):186-93.

[66] El-Hassan H, Shao Y, Ghouleh Z. Reaction Products in Carbonation-Cured Lightweight Concrete. J Mater Civ Eng. 2013;25(6):799-809.

[67] Jang JG, Lee HK. Microstructural densification and CO2 uptake promoted by the carbonation curing of belite-rich Portland cement. Cem Concr Res. 2016;82:50-7.

[68] Shah V, Scrivener K, Bhattacharjee B, Bishnoi S. Changes in microstructure characteristics of cement paste on carbonation. Cem Concr Res. 2018;109:184-97.

[69] Soroushian P, Won J-P, Hassan M. Durability characteristics of CO2-cured cellulose fiber reinforced cement

composites. Constr Build Mater. 2012;34:44-53.

[70] Rostami V, Shao, Y., and Boyd, A. J. Durability of concrete pipes subjected to combined steam and carbonation curing. Constr Build Mater. 2011;25:3345-55.

[71] Goto S, Suenaga, K., and Kado, T. Calcium silicate carbonation products. Journal of the American Ceramic Society. 1995;78(11):2867-72.

[72] Castellote M, and Fernandez, L. Chemical changes and phase analysis of OPC pastes carbonated at different CO2 concentrations. Mater Struct. 2009;42(4):515-25.

[73] El-Hassan H, Shao Y. Dynamic carbonation curing of fresh lightweight concrete. Mag Concr Res. 2014;66(14):708-18.

[74] Goodbrake CJ, Young, J. F., and Berger, R. L. Reaction of hydraulic calcium silicates with carbon dioxide and water. Journal of the American Ceramic Society. 1979;62(9-10):488-91.

[75] Berger RL. Stabilization of silicate structures by carbonation. Cem Concr Res. 1979;9(5):649-51.

[76] Ramezanianpour AA, Ghahari SA, Esmaeili M. Effect of combined carbonation and chloride ion ingress by an accelerated test method on microscopic and mechanical properties of concrete. Constr Build Mater. 2014;58:138-46.

[77] Liu Y, Zhuge Y, Chow CWK, Keegan A, Li D, Pham PN, et al. Properties and microstructure of concrete blocks incorporating drinking water treatment sludge exposed to early-age carbonation curing. J Clean Prod. 2020;261:121257.

[78] Groves GW, and Brough, A. Progressive Changes in the Structure of Hardened C3S Cement Pastes Due to Carbonation. Journal of the American Ceramic Society. 1991;74(11):2891-6.

[79] Mo L, Panesar DK. Effects of accelerated carbonation on the microstructure of Portland cement pastes containing reactive MgO. Cem Concr Res. 2012;42(6):769-77.

[80] El-Hassan H, Shao Y. Early carbonation curing of concrete masonry units with Portland limestone cement. Cem Concr Compos. 2015;62:168-77.

[81] Zhang D, Cai X, Jaworska B. Effect of pre-carbonation hydration on long-term hydration of carbonation-cured cement-based materials. Constr Build Mater. 2020;231:117122.

[82] El-Hassan H, Shao Y, Ghouleh Z. Effect of Initial Curing on Carbonation of Lightweight Concrete Masonry Units. ACI Mater J. 2013;110(4):441-50.

[83] Mo L, Zhang F, Deng M. Mechanical performance and microstructure of the calcium carbonate binders produced by carbonating steel slag paste under CO2 curing. Cem Concr Res. 2016;88:217-26.

[84] He P, Shi C, Tu Z, Poon CS, Zhang J. Effect of further water curing on compressive strength and microstructure of CO2-cured concrete. Cem Concr Compos. 2016;72:80-8.

[85] Zhang D, Cai X, Shao Y. Carbonation Curing of Precast Fly Ash Concrete. J Mater Civ Eng. 2016;28(11):04016127.

[86] Shi C, He F, Wu Y. Effect of pre-conditioning on CO2 curing of lightweight concrete blocks mixtures. Constr Build Mater. 2012;26(1):257-67.

[87] El-Hassan H. Static and Dynamic Carbonation of Lightweight Concrete Masonry Units. Montreal, QC: McGill University; 2012.

[88] Behfarnia K, Rostami M. An assessment on parameters affecting the carbonation of alkali-activated slag concrete. J Clean Prod. 2017;157:1-9.

[89] Monkman S, MacDonald M. Carbon dioxide upcycling into industrially produced concrete blocks. Constr Build Mater. 2016;124:127-32.

[90] Zhang D, Shao Y. Early age carbonation curing for precast reinforced concretes. Constr Build Mater. 2016;113:134-43.

[91] Matsushita F, Aono, Y., and Shibata, S. Carbonation degree of autoclaved aerated concrete. Cem Concr Res. 2000;30:1741-5.

[92] El-Hassan H, Shao Y. Carbonation Curing of Concrete Blocks to Mitigate Carbon Emission. The Conference on Carbon Capture, Utilization, and Storage, Annual CCUS 2015 Conference; Pittsburgh, PA, USA.2015.

[93] El-Hassan H, Shao Y. Carbon Storage through Concrete Block Carbonation. Journal of Clean Energy Technologies. 2014;2(3):287-91.

[94] El-Hassan H, Shao Y. Carbon Storage through Dynamic Carbonation of Fresh Lightweight Concrete. American Institute of Chemical Engineers (AIChE) 2016 Spring Meeting; Houston, TX, USA2016.

[95] He Z, Wang S, Mahoutian M, Shao Y. Flue gas carbonation of cement-based building products. Journal of CO2 Utilization. 2020;37:309-19.

[96] Neves Junior A, Toledo Filho RD, de Moraes Rego Fairbairn E, Dweck J. The effects of the early carbonation curing on the mechanical and porosity properties of high initial strength Portland cement pastes. Constr Build Mater. 2015;77:448-54.

[97] Xuan D, Zhan B, Poon CS. Development of a new generation of eco-friendly concrete blocks by accelerated mineral carbonation. J Clean Prod. 2016;133:1235-41.

[98] Ahmad S, Assaggaf RA, Maslehuddin M, Al-Amoudi OSB, Adekunle SK, Ali SI. Effects of carbonation pressure and duration on strength evolution of concrete subjected to accelerated carbonation curing. Constr Build Mater. 2017;136:565-73.

[99] Monkman S, Shao Y. Carbonation Curing of Slag-Cement Concrete for Binding CO2 and Improving Performance. J Mater Civ Eng. 2010;22(4):296-304.

[100] Monkman S, Shao Y. Assessing the carbonation behavior of cementitious materials. J Mater Civ Eng. 2006;November/December:768-76.

[101] Rostami V, Shao Y, Boyd AJ. Carbonation Curing versus Steam Curing for Precast Concrete Production. J Mater Civ Eng. 2012;24(9):1221-9.

[102] Wang L, Yeung TLK, Lau AYT, Tsang DCW, Poon C-S. Recycling contaminated sediment into eco-friendly paving blocks by a combination of binary cement and carbon dioxide curing. J Clean Prod. 2017;164:1279-88.

[103] Scrivener K, Snellings R, Lothenbach B. A Practical Guide to Microstructural Analysis of Cementitious Materials: CRC Press; 2015.

[104] Ramachandran VS, Beaudoin JJ. Handbook of Analytical Techniques in Concrete Science and Technology: William Andre Publishing/Noyes; 2001.

[105] Bukowski JM, and Berger, R. L. Reactivity and strength development of CO2 activated non-hydraulic calcium silicates. Cem Concr Res. 1979;9(1):57-68.

[106] Chang C, and Chen, J. The experimental investigation of concrete carbonation depth. Cem Concr Res. 2006;36:1760-7.

[107] Huffman EWD. Performance of a new automatic carbon dioxide coulometer. Microchemical Journal. 1977;22(4):567-73.

[108] El-Hassan H, Shao Y. Innovative CO2 Utilization By Carbonation Curing of Lightweight Concrete Made with Portland Limestone Cement. American Institute of Chemical Engineers (AIChE) 2016 Annual Meeting; San Francisco, CA, USA.2016.

[109] Shao Y, El-Hassan H, editors. CO2 Utilization in Concrete Block Production. American Institute of Chemical Engineers (AIChE) Annual Meeting 2012; 2012; Pittsburgh, PA, USA.

[110] Shao Y, El-Hassan H. CO2 Utilization in Concrete. Third International Conference on Sustainable Construction Materials and Technologies (SCMT3), ; Kyoto, Japan2013.

[111] Liu Y, Zhuge Y, Chow CWK, Keegan A, Pham PN, Li D, et al. Recycling drinking water treatment sludge into eco-concrete blocks with CO2 curing: Durability and leachability. Sci Total Environ. 2020;746:141182.

[112] Zhang S, Ghouleh Z, Shao Y. Effect of Carbonation Curing on Efflorescence Formation in Concrete Paver Blocks. J Mater Civ Eng. 2020;32(6):04020127.

[113] Shao Y, Monkman S, Wang S. Market Analysis of CO2 Sequestration in Concrete Building Products. Second International Conference on Sustainable Construction Materials and Technologies; Acona, Italy2010.

[114] Shao Y, Lin X. Early-Age Carbonation Curing of Concrete Using Recovered CO2. Concrete International. 2011;33(9):50-6.

[115] Zhang D, Shao Y. Enhancing Chloride Corrosion Resistance of Precast Reinforced Concrete by Carbonation Curing. ACI Mater J. 2019;116(3):3-12.

[116] Shao Y, Wang S. Carbonation Curing of Cement Bonded Fiberboard Made by Slurry-Dewatering Process. ACI Symposium Publication. 2009;260:125-38.

[117] He Z, Jia Y, Wang S, Mahoutian M, Shao Y. Maximizing CO2 sequestration in cement-bonded fiberboards through carbonation curing. Constr Build Mater. 2019;213:51-60.

[118] Maail RS, Umemura K, Aizawa H, Kawai S. Curing and degradation processes of cement-bonded particleboard by supercritical CO2 treatment. Journal of Wood Science. 2011;57(4):302-7.

[119] Maail RS. Degradation Analysis on Manufacture of Cement-bonded Particleboard Using Supercritical CO2. 2017.

[120] Soroushian P, Aouadi F, Chowdhury H, Nossoni A, Sarwar G. Cement-bonded straw board subjected to accelerated processing. Cem Concr Compos. 2004;26(7):797-802.

[121] Wang L, Chen SS, Tsang DCW, Poon C-S, Dai J-G. CO2 curing and fibre reinforcement for green recycling of contaminated wood into high-performance cement-bonded particleboards. Journal of CO2 Utilization. 2017;18:107-16.

[122] Soroushian P, Hassan M. Evaluation of cement-bonded strawboard against alternative cement-based siding products. Constr Build Mater. 2012;34:77-82.

[123] Wang L, Chen SS, Tsang DCW, Poon C-S, Shih K. Recycling contaminated wood into eco-friendly particleboard using green cement and carbon dioxide curing. J Clean Prod. 2016;137:861-70.

[124] Martín D, Aparicio P, Galán E. Accelerated carbonation of ceramic materials. Application to bricks from Andalusian factories (Spain). Constr Build Mater. 2018;181:598-608.

[125] Transparency Market Research. Global Concrete Block and Brick Manufacturing Market Estimated to Surpass 2700 Billion Units by 2027. Albany, NY, USA: Transparency Market Research, 2018.

[126] Concrete Manufacturers Association. Concrete Block Paving. 5th Edition ed. Midrand, South Africa: Concrete Manufacturers Association; 2009. 30 p.

[127] Lucintel. Concrete Pipe Market Report: Trends, Forecast and Competitive Analysis. Dallas, TX, USA: 2020.

[128] Railway Tie Association. https://www.rta.org/faqs#:~:text=Currently%2C%20the%20industry%20has%20the,just%20over%2019%20million%20ties. Georgia, USA: Railway Tie Association; 2020 [cited 2020].

[129] Saunders A, Davidson E. Cement Boards. Global Cement. 2014:32-8.

[130] ACIMAC. World Production and Consumption of Ceramic Tiles. Italy: Association of Italian Manufacturers of Machinery and Equipment for Ceramics, 2018.

[131] Cement production worldwide from 1995 to 2019. https://www.statista.com/statistics/1087115/global-cement-production-volume/ [cited 2020].

[132] El-Dieb AS. Self-curing concrete: Water retention, hydration and moisture transport. Constr Build Mater. 2007;21(6):1282-7.

Chapter 9

Cement-Based Piezoelectricity Application: A Theoretical Approach

Daniel A. Triana-Camacho, Jorge H. Quintero-Orozco and Jaime A. Perez-Taborda

Abstract

The linear theory of piezoelectricity has widely been used to evaluate the material constants of single crystals and ceramics, but what happens with amorphous structures that exhibit piezoelectric properties such as cement-based? In this chapter, we correlate the theoretical and experimental piezoelectric parameters for small deformations after compressive stress–strain, open circuit potential, and impedance spectroscopy on cement-based. Here, in detail, we introduce the theory of piezoelectricity for large deformations without including a functional for the energy; also, we show two generating equations in terms of a free energy's function for later it will be reduced to constitutional equations of piezoelectricity for infinitesimal deformations. Finally, here is shown piezoelectric and electrical parameters of gold nanoparticles mixed to cement paste: the axial elasticity parameter $Y = 323.5 \pm 75.3$ $[kN/m^2]$, the electroelastic parameter $\gamma = -20.5 \pm 6.9$ $[mV/kN]$, and dielectric constant $\varepsilon = (939.6 \pm 82.9)\varepsilon_0$ $[F/m]$, which have an interpretation as linear theory parameters s^D_{ijkl}, g_{kij} and ε^T_{ik} discussed in the chapter.

Keywords: piezoelectricity, cement-based, nano-composites, constitutional equations, impedance spectroscopy

1. Introduction

The direct piezoelectric effect creates an electric polarization on a continuum medium due to applied stress. The polarization can be macroscopic (effect over continuum medium) and nanoscopic and microscopy scales (effect over atoms, molecules, and electrical domains). Once the Curie brothers discovered the piezoelectric effect in 1880 [1], piezoelectricity investigations led to more data and constructed models based on crystallography to explain the electricity generation since electro-optics and thermodynamic. Voigt in 1894 proposed a piezoelectric parameter related to the strain of material; since the thermodynamic theory, he constructed a non-linear model and expressed the free energy of a piezoelectric crystal in terms of the electric field, strain, electric and elastic deformation potentials, temperature, pyroelectric and piezoelectric parameters [2]. Currently, we can see these constants in the constitutive equations of piezoelectricity. During 1956 and 1963, Toupin and Eringen used a variational formulation to construct a functional in terms of internal energy and derive the constitutive equations [3, 4]. Then, in 1971 Tiersten proposed to use the conservation equations of mass, electrical charge, linear momentum, angular momentum, and energy, adding a Legendre transformation to include a thermodynamic functional in terms of the free energy, achieving a reduction of the number of constitutive equations from 7 to 4 to facilitate theoretical calculations [5]. These constitutional equations and their linear approach gave support to the theoretical calculus of piezoelectric parameters of crystalline structures, e.g., zinc-blende [6, 7], zinc oxide [8, 9], and other crystals with similar symmetric of quartz [10, 11]. Finally, between 1991 and 2017, Yang has proposed modifications for the Legendre transformation of Tiersten, and he has included two models to describe the polarization in a deformable continuum medium [12, 13].

According to electrostatic theory, the macroscopic polarization $\vec{P}$ can be written in terms of electric charge distribution likewise with the electric field $\vec{E}$ induced into the continuum medium. Besides, the polarization starts with continuum medium deformation $\vec{S}$ for applied stress. Considering: (i) Uniqueness for the parameters that relate the polarization and the electric field, (ii) Stress produces equal deformations in each cell of crystal, (iii) The deformations lead dipole and quadrupole moments affecting the piezoelectric parameters directly. Based on the above considerations, the linear constitutive equations of piezoelectricity can be written, as Martin said [6].

This chapter book is thought to be a working example that connects the piezoelectricity theory and experimental data of electromechanical and electrical properties. These data were obtained on cement paste mixed with gold nanoparticles.

2. Constitutional equations in detail and correlation with the piezoelectricity of cement-based composites

In this section, we have selected Yang's differential approach to obtain the constitutional equations of piezoelectricity. The differential derivation shows the physics involved in the conservation laws differently. For example, it shows that the electric body couple and the Cauchy stress tensor are asymmetric. Also, it relates the local electric field with the electric interaction between the differential elements of the lattice continuum and the electronic continuum.

2.1 Conservation laws applied to a polarized continuum from differential approximation

Regarding the study of the piezoelectric properties of cement paste, it is necessary to describe the material separately as two continua medium, as from the piezoelectric phenomenon, the crystal and their symmetry would have a lattice (positive charge) and an electronic component (negative charge) Those continua can be separated by mechanical stress. Their physical properties could change according to the coordinate systems or states. Therefore, the body will study in two states (reference state and current state), as is shown in **Figure 1**.

2.1.1 Electric charge conservation

The conservation electric charge in the body takes importance with an infinitesimal displacement on the medium's current state to get polarization. For this reason, the phenomenon described in the above state is known as the two continuum medium model. The electronic continuum comes under an infinitesimal

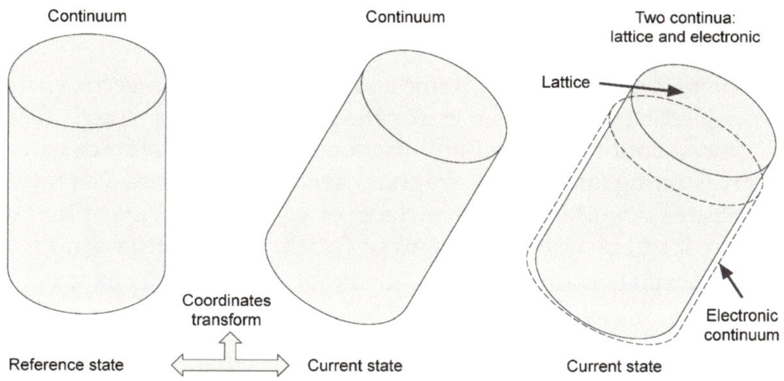

Figure 1.
States of a deformable and polarizable continuum.

displacement η respect to the lattice. It is produced by body deformation, as described in **Figure 2**. From the assumption of the lattice and electronic continuum, the medium have equal volumes, some variation of infinitesimal displacement respect coordinates in the current state should be zero and can be written as

$$\eta_{k,k} = \frac{\partial \eta_k(y)}{\partial y_k} = 0 \tag{1}$$

Furthermore, if it is taken the two continuous mediums, the electric charge density must be neutral to consider only the piezoelectric effect.

$$\mu^l\left(\vec{y}\right) + \mu^e\left(\vec{y} + \vec{\eta}\right) = 0 \tag{2}$$

We can show that the gradient of infinitesimal displacement Eq. (1) and the neutrality condition of electric charge density Eq. (2) are sufficient to explain the polarization in a deformable continuum

$$\vec{P} = \mu^l\left(\vec{y}\right)\left[-\vec{\eta}\left(\vec{y}\right)\right] = \mu^e\left(\vec{y} + \vec{\eta}\right)\vec{\eta} \approx \mu^e\left(\vec{y}\right)\vec{\eta} \tag{3}$$

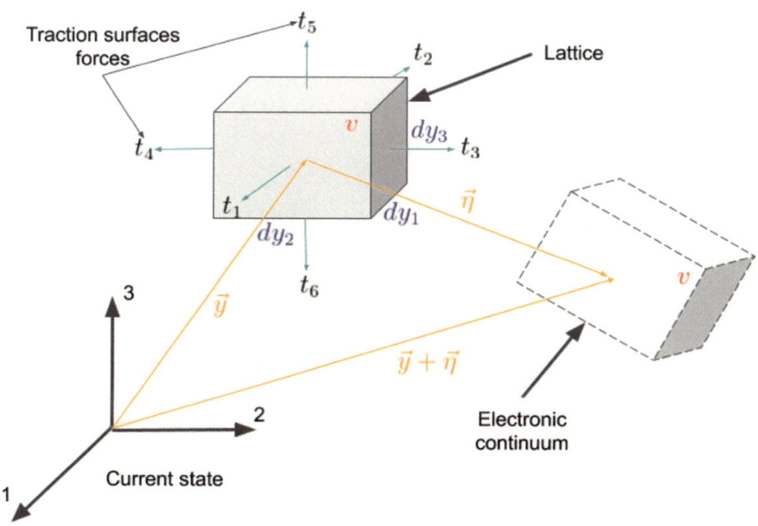

Figure 2.
Volume elements of electronic and lattice continuum medium.

2.1.2 Energy conservation

Once the body is deformed, electronic and lattice continua electric charges apply a quasi-static electric field. Tiersten et al. called it Maxwelliam electric field E_k. It is interacting on two continuum mediums producing an electrical force on each one. The other forces acting on the body are traction and body forces. The traction force t_k per unit, the area is working on the surfaces of volume elements of the lattice (see **Figure 2**). Also, it can be written in terms of Cauchy stress tensor τ_{jk} as $t_k = n_j\tau_{jk}$, where n_j is the normal vector. Moreover, body force refers to an external force acting on the body, for example, gravity. The previous three forces are necessary to get the conservation laws, including energy.

From the three forces above, it is possible to construct the Eqs. (4) and (5), linear and angular momentum conservation, respectively.

$$\rho\dot{u}_k = f_k^E + \tau_{mk,m} + \rho f_k \tag{4}$$

$$c_i^E + \varepsilon_{ijk}\tau_{jk}\left(\vec{y}\right) = 0 \tag{5}$$

where f_k^E is the electric force on the body and start from the dot product of polarization P_j with the electric field gradient $E_{k,i}$; Besides, c_i^E is called electric coupling, and it is the cross product of P_j with E_k; finally, f_k is the external force on the body per unit mass.

Then, replacing the electric coupling c_i^E by the cross product between P_j and E_k, we can rewrite Eq. (5) as follow

$$\varepsilon_{ijk}P_jE_k\left(\vec{y}\right) + \varepsilon_{ijk}\tau_{jk}\left(\vec{y}\right) = 0 \tag{6}$$

Factoring the permutation tensor ε_{ijk}

$$\varepsilon_{ijk}\left[P_jE_k + \tau_{jk}\right] = 0 \tag{7}$$

The term in square brackets from Eq. (7) is a symmetry tensor that can be written as

$$\tau^S{}_{jk} = P_jE_k + \tau_{jk} \tag{8}$$

The term P_jE_k is called the Maxwell stress tensor $T^E{}_{jk}$. On the other hand, multiplying Eq. (5) by ε_{iqr} is obtain

$$\begin{gathered}\varepsilon_{iqr}c_i^E + \varepsilon_{iqr}\varepsilon_{ijk}\tau_{jk}\left(\vec{y}\right) = 0\\ \delta_{jq}\delta_{kr}\tau_{jk}\left(\vec{y}\right) - \delta_{jr}\delta_{kq}\tau_{jk}\left(\vec{y}\right) = -\varepsilon_{iqr}c_i^E\\ \tau_{qr}\left(\vec{y}\right) - \tau_{rq}\left(\vec{y}\right) = -\varepsilon_{iqr}c_i^E\end{gathered} \tag{9}$$

From Eq. (9) we can conclude that the Cauchy stress tensor ($t_k = n_j\tau_{jk}$) is asymmetry. Now, we get the total energy inside the continuum medium as a combination of kinetic and internal energy ε^{in} [13], both per mass unit. Here is performed the powers added due to the three forces above.

$$\frac{d}{dt}\left(\frac{1}{2}u_k u_k \rho dv + \varepsilon^{in}\rho dv\right) = \mu^l\left(\vec{y}\right)E_k\left(\vec{y}\right)u_k\left(\vec{y}\right)dv + \mu^e\left(\vec{y}+\vec{\eta}\right)E_k\left(\vec{y}+\vec{\eta}\right)\left[u_k\left(\vec{y}\right)+\dot{\eta}_k\right]dv + t_k u_k ds + \rho f_k u_k dv \tag{10}$$

The following steps from Eq. (10) conduct to develop conservation energy law. **In the first step**, we work on the **total energy term**. Then, the product rule to the total energy term is applied:

$$\begin{aligned}\frac{d}{dt}\left(\frac{1}{2}u_k u_k \rho dv + \varepsilon^{in}\rho dv\right) &= \rho dv\frac{d}{dt}\left(\frac{1}{2}u_k u_k + \varepsilon^{in}\right) + \left(\frac{1}{2}u_k u_k + \varepsilon^{in}\right)\frac{d}{dt}(\rho dv)\\ &= \rho dv\frac{d}{dt}\left(\frac{1}{2}u_k u_k + \varepsilon^{in}\right) + \left(\frac{1}{2}u_k u_k + \varepsilon^{in}\right)\frac{d}{dt}(\rho dv)\end{aligned} \tag{11}$$

The term $\frac{d}{dt}(\rho dv)$ represents the mass conservation. Therefore, this term is null in Eq. (11).

$$\frac{d}{dt}\left(\frac{1}{2}u_k u_k \rho dv + \varepsilon^{in}\rho dv\right) = \rho dv\frac{d}{dt}\left(\frac{1}{2}u_k u_k + \varepsilon^{in}\right) \tag{12}$$

Then, here is writes the add of derivatives from kinetic and internal energies

$$\frac{d}{dt}\left(\frac{1}{2}u_k u_k \rho dv + \varepsilon^{in}\rho dv\right) = \rho dv\, u_k\frac{d}{dt}(u_k) + \rho dv\frac{d}{dt}\left(\varepsilon^{in}\right) \tag{13}$$

In the second step, the **electric power term** will be developed by Taylor's expansion.

$$\begin{aligned}&\mu^l\left(\vec{y}\right)E_k\left(\vec{y}\right)u_k\left(\vec{y}\right)dv + \mu^e\left(\vec{y}+\vec{\eta}\right)E_k\left(\vec{y}+\vec{\eta}\right)\left[u_k\left(\vec{y}\right)+\dot{\eta}_k\right]dv\\ &\approx \mu^l\left(\vec{y}\right)E_k\left(\vec{y}\right)u_k\left(\vec{y}\right)dv + \mu^e\left(\vec{y}+\vec{\eta}\right)\left[E_k\left(\vec{y}\right)+E_{k,i}\left(\vec{y}\right)\eta_i\right]\left[u_k\left(\vec{y}\right)+\dot{\eta}_k\right]dv\end{aligned} \tag{14}$$

From Eq. (14), it will solve the dot product between the electric field and velocity,

$$\begin{aligned}&= \mu^l\left(\vec{y}\right)E_k\left(\vec{y}\right)u_k\left(\vec{y}\right)dv\\ &+\mu^e\left(\vec{y}+\vec{\eta}\right)\left[E_k\left(\vec{y}\right)u_k\left(\vec{y}\right) + E_k\left(\vec{y}\right)\dot{\eta}_k + E_{k,i}\left(\vec{y}\right)\eta_i u_k\left(\vec{y}\right) + E_{k,i}\left(\vec{y}\right)\eta_i\dot{\eta}_k\right]dv\end{aligned} \tag{15}$$

The second-order term $\eta_i\dot{\eta}_k = \eta_i\dot{\eta}_i = \frac{1}{2}\frac{d}{dt}\eta_i{}^2 \approx 0$ is zero, taking into account the infinitesimal displacement. Then, factorize $E_k\left(\vec{y}\right)u_k\left(\vec{y}\right)dv$ in Eq. (15), it takes a form:

$$= \left[\mu^l\left(\vec{y}\right) + \mu^e\left(\vec{y}+\vec{\eta}\right)\right]E_k\left(\vec{y}\right)u_k\left(\vec{y}\right)dv + \left[E_k\left(\vec{y}\right)\dot{\eta}_k + E_{k,i}\left(\vec{y}\right)\eta_i u_k\left(\vec{y}\right)\right]dv \tag{16}$$

Replacing Eq. (2) in Eq. (16), we obtain:

$$= \mu^e\left(\vec{y}+\vec{\eta}\right)E_k\left(\vec{y}\right)\dot{\eta}_k dv + \mu^e\left(\vec{y}+\vec{\eta}\right)E_{k,i}\left(\vec{y}\right)\eta_i u_k\left(\vec{y}\right)dv \tag{17}$$

With Eq. (3), Eq. (17) takes the form:

$$= \mu^e\left(\vec{y}+\vec{\eta}\right)E_k\left(\vec{y}\right)\dot{\eta}_k dv + P_i E_{k,i}\left(\vec{y}\right)u_k\left(\vec{y}\right)dv \tag{18}$$

In Eq. (18), the term $\mu^e\left(\vec{y}+\vec{\eta}\right)E_k\left(\vec{y}\right)\dot{\eta}_k$ is called the electric power of the body w^E And the term $P_i E_{k,i}$ is the electric force between the lattice and electronic volume elements. It was defined in Eq. (4) above.

$$\mu^l\left(\vec{y}\right)E_k\left(\vec{y}\right)u_k\left(\vec{y}\right)dv + \mu^e\left(\vec{y}+\vec{\eta}\right)E_k\left(\vec{y}+\vec{\eta}\right)\left[u_k\left(\vec{y}\right)+\dot{\eta}_k\right]dv = w^E dv + f_k^{\ E} u_k\left(\vec{y}\right)dv \tag{19}$$

In the third step, we will solve the **traction force** in terms of Cauchy stress tensor. Then, the power due to the traction force takes the form:

$$\begin{aligned} t_k u_k ds &= \tau_{mk} n_m u_k ds \\ &= \tau_{1k}\left(\vec{y}+\frac{1}{2}dy_1\hat{i}_1\right)u_k\left(\vec{y}+\frac{1}{2}dy_1\hat{i}_1\right)dy_2 dy_3 \\ &\quad -\tau_{1k}\left(\vec{y}-\frac{1}{2}dy_1\hat{i}_1\right)u_k\left(\vec{y}-\frac{1}{2}dy_1\hat{i}_1\right)dy_2 dy_3 \\ &\quad +\tau_{2k}\left(\vec{y}+\frac{1}{2}dy_2\hat{i}_2\right)u_k\left(\vec{y}+\frac{1}{2}dy_2\hat{i}_2\right)dy_3 dy_1 \\ &\quad -\tau_{2k}\left(\vec{y}-\frac{1}{2}dy_2\hat{i}_2\right)u_k\left(\vec{y}-\frac{1}{2}dy_2\hat{i}_2\right)dy_3 dy_1 \\ &\quad +\tau_{3k}\left(\vec{y}+\frac{1}{2}dy_3\hat{i}_3\right)u_k\left(\vec{y}+\frac{1}{2}dy_3\hat{i}_3\right)dy_1 dy_2 \\ &\quad -\tau_{3k}\left(\vec{y}-\frac{1}{2}dy_3\hat{i}_3\right)u_k\left(\vec{y}-\frac{1}{2}dy_3\hat{i}_3\right)dy_1 dy_2 \end{aligned} \tag{20}$$

Now, we apply Taylor's expansion to Cauchy stress tensor τ_{1k} and velocity u_k. Next, the last result will be implemented in the components: two τ_{2k} and three τ_{3k}

$$\begin{aligned} &\tau_{1k}\left(\vec{y}+\frac{1}{2}dy_1\hat{i}_1\right)u_k\left(\vec{y}+\frac{1}{2}dy_1\hat{i}_1\right)dy_2 dy_3 - \tau_{1k}\left(\vec{y}-\frac{1}{2}dy_1\hat{i}_1\right)u_k\left(\vec{y}-\frac{1}{2}dy_1\hat{i}_1\right)dy_2 dy_3 \\ &\approx \left[\tau_{1k}\left(\vec{y}\right)+\frac{1}{2}dy_1\tau_{1k,1}\left(\vec{y}\right)\right]\left[u_k\left(\vec{y}\right)+\frac{1}{2}dy_1 u_{k,1}\left(\vec{y}\right)\right]dy_2 dy_3 \\ &\quad -\left[\tau_{1k}\left(\vec{y}\right)-\frac{1}{2}dy_1\tau_{1k,1}\left(\vec{y}\right)\right]\left[u_k\left(\vec{y}\right)-\frac{1}{2}dy_1 u_{k,1}\left(\vec{y}\right)\right]dy_2 dy_3 \end{aligned} \tag{21}$$

The products give

$$\approx\left[\tau_{1k}u_k+\tau_{1k}\frac{1}{2}dy_1u_{k,1}+\frac{1}{2}dy_1\tau_{1k,1}u_k+\frac{1}{4}dy_1{}^2\tau_{1k,1}u_{k,1}\right]dy_2dy_3$$
$$-\left[\tau_{1k}u_k-\tau_{1k}\frac{1}{2}dy_1u_{k,1}-\frac{1}{2}dy_1\tau_{1k,1}u_k+\frac{1}{4}dy_1{}^2\tau_{1k,1}u_{k,1}\right]dy_2dy_3 \tag{22}$$

Adding similar terms:

$$\approx\left[\tau_{1k}\left(\vec{y}\right)dy_1u_{k,1}\left(\vec{y}\right)+dy_1\tau_{1k,1}\left(\vec{y}\right)u_k\left(\vec{y}\right)\right]dy_2dy_3 \tag{23}$$

Factorizing dy_1 that multiply to dy_2 and dy_3 it will transform into the volume element dv.

$$\approx\tau_{1k}\left(\vec{y}\right)u_{k,1}\left(\vec{y}\right)dv+\tau_{1k,1}\left(\vec{y}\right)u_k\left(\vec{y}\right)dv \tag{24}$$

Reply the proceeding since Eq. (21) to Eq. (24) for the components τ_{2k} and τ_{3k} yield

$$t_ku_kds=\tau_{1k}\left(\vec{y}\right)u_{k,1}\left(\vec{y}\right)dv+\tau_{1k,1}\left(\vec{y}\right)u_k\left(\vec{y}\right)dv+\tau_{2k}\left(\vec{y}\right)u_{k,2}\left(\vec{y}\right)dv$$
$$+\tau_{2k,2}\left(\vec{y}\right)u_k\left(\vec{y}\right)dv+\tau_{3k}\left(\vec{y}\right)u_{k,3}\left(\vec{y}\right)dv+\tau_{3k,3}\left(\vec{y}\right)u_k\left(\vec{y}\right)dv \tag{25}$$

Each one of the components has information about two opposite faces of the volume element. Then, adding index m it reduces the Eq. (25) to:

$$t_ku_kds=\tau_{mk}\left(\vec{y}\right)u_{k,m}\left(\vec{y}\right)dv+\tau_{mk,m}\left(\vec{y}\right)u_k\left(\vec{y}\right)dv \tag{26}$$

From the results of Eq. (13), Eq. (19), and Eq. (26) into Eq. (10), we obtain

$$\rho dv\,u_k\frac{d}{dt}(u_k)+\rho dv\frac{d}{dt}\left(\varepsilon^{in}\right)=w^Edv+f_k{}^Eu_k\left(\vec{y}\right)dv+\tau_{mk}\left(\vec{y}\right)u_{k,m}\left(\vec{y}\right)dv$$
$$+\tau_{mk,m}\left(\vec{y}\right)u_k\left(\vec{y}\right)dv+\rho f_ku_kdv \tag{27}$$

Here is factoring the terms that contain $u_k\left(\vec{y}\right)dv$ to the left side and the terms with the volume differential dv to the right side.

$$\left[\rho\frac{d}{dt}(u_k)+f_k{}^E+\tau_{mk,m}\left(\vec{y}\right)+\rho f_k\right]u_k\left(\vec{y}\right)dv=\left\{w^E-\rho\frac{d}{dt}\left[\varepsilon^{in}\right]+\tau_{mk}\left(\vec{y}\right)u_{k,m}\left(\vec{y}\right)\right\}dv \tag{28}$$

From Eq. (28), the term in the square bracket is null by Eq. (4). Then, we obtain Eq. (29) for energy conservation that depends on internal energy.

$$\rho\dot{\varepsilon}^{in}=w^E+\tau_{mk}\left(\vec{y}\right)u_{k,m}\left(\vec{y}\right) \tag{29}$$

Remember from Eq. (19) that electric power can be written as:

$$w^E=\mu^e\left(\vec{y}+\vec{\eta}\right)E_m\left(\vec{y}\right)\frac{d\eta_m}{dt}=E_m\left(\vec{y}\right)\left\{\frac{d}{dt}\left[\mu^e\left(\vec{y}+\vec{\eta}\right)\eta_m\right]-\frac{d\mu^e\left(\vec{y}+\vec{\eta}\right)}{dt}\eta_m\right\}$$

$$w^E = E_m\left(\vec{y}\right)\left\{\dot{P}_m - \dot{\mu}^e\left(\vec{y} + \vec{\eta}\right)\eta_m\right\} \tag{30}$$

Another form of electric charge conservation is $\dot{\mu}^e\left(\vec{y} + \vec{\eta}\right) + \mu^e\left(\vec{y} + \vec{\eta}\right)u_{i,i} = 0$, it will simplify the Eq. (30) to:

$$w^E = E_m\left(\vec{y}\right)\left\{\dot{P}_m + \mu^e\left(\vec{y} + \vec{\eta}\right)u_{i,i}\eta_m\right\} \tag{31}$$

The mass conservation $\dot{\rho} + \rho u_{i,i} = 0$ has a similar mathematical structure as charge conservation. Therefore, the gradient of the speed $u_{i,i}$ in Eq. (31) was replaced

$$w^E = E_m\left(\vec{y}\right)\dot{P}_m + E_m\left(\vec{y}\right)\mu^e\left(\vec{y} + \vec{\eta}\right)\frac{-\dot{\rho}}{\rho}\eta_m = E_m\dot{P}_m - \frac{\dot{\rho}}{\rho}E_m P_m$$

$$w^E = \frac{E_m}{\rho}\left[\rho\dot{P}_m - \dot{\rho}P_m\right] \tag{32}$$

Eq. (32) has been used on Eq. (29)

$$\rho\dot{\varepsilon}^{in} = \frac{E_m}{\rho}\left[\rho\dot{P}_m - \dot{\rho}P_m\right] + \tau_{mk}\left(\vec{y}\right)u_{k,m}\left(\vec{y}\right) \tag{33}$$

With Legendre transformation showing in Eq. (34), Tiersten replaced the internal energy ε^{in} by free energy χ [5]. This transformation diminishes the number of constitutional equations. Besides, it offers a quantitative interpretation that can not get from the internal energy resulting in more useful for those who perform piezoelectricity experiments. After Section 2.3, we could see the χ will depend on the gradient of potential in the reference state and deformation tensor.

$$\chi = \varepsilon^{in} - E_m\frac{P_m}{\rho} \tag{34}$$

Upon differentiating respect to time the Eq. (34).

$$\dot{\chi} = \dot{\varepsilon}^{in} - \dot{E}_m\frac{P_m}{\rho} - E_m\frac{\dot{P}_m}{\rho} + E_m\frac{P_m}{\rho^2}\dot{\rho} \tag{35}$$

Clear the term $\rho\dot{\varepsilon}^{in}$

$$\rho\dot{\varepsilon}^{in} = \rho\dot{\chi} + \dot{E}_m P_m + E_m\dot{P}_m - E_m\frac{P_m}{\rho}\dot{\rho} \tag{36}$$

Using Eq. (36) on Eq. (29), we obtain:

$$\rho\dot{\chi} + \dot{E}_m P_m + E_m\dot{P}_m - E_m\frac{P_m}{\rho}\dot{\rho} = \frac{E_m}{\rho}\left[\rho\dot{P}_m - \dot{\rho}P_m\right] + \tau_{mk}\left(\vec{y}\right)u_{k,m}\left(\vec{y}\right)$$

$$\rho\dot{\chi} + \dot{E}_m P_m + E_m\dot{P}_m - E_m\dot{\rho}\frac{P_m}{\rho} = E_m\dot{P}_m - E_m\dot{\rho}\frac{P_m}{\rho} + \tau_{mk}\left(\vec{y}\right)u_{k,m}\left(\vec{y}\right) \tag{37}$$

The similar terms $E_m\dot{P}_m$ and $-E_m\dot{\rho}\frac{P_m}{\rho}$ in Eq. (37), are clear. Finally, we have rewritten **energy conservation** in terms of free energy, electrical (electric field and polarization vector), and mechanical (Cauchy stress tensor) components, as is shown in Eq. (38).

$$\rho\dot{\chi} = \tau_{mk}\left(\vec{y}\right)u_{k,m}\left(\vec{y}\right) - \dot{E}_m P_m \tag{38}$$

2.2 Transformation of fundamental physical quantities in piezoelectricity to the reference state

There are several reasons to consider two coordinate systems (reference and current state) for continuum. Firstly, it is not mathematically simple to describe the movement of each particle that compounds a continuum as seen on the gradient of velocity $u_{k,m}$ in Eq. (38); it is more appropriate to propose a coordinate system that describes the continuum in the reference system. The material behavior could be affected by the characteristics of the current state, too. For example, fluids and solids can change their mechanical behavior while changing the shape [14]. Hence, we refer to our study material (cement-based composites) whom we know the physical properties in the reference state X_L. To explain the behavior of a material, we must include physical quantities respect the reference state X_L: potential gradient W_K, polarization P_L, electric displacement $\mathcal{D}_L$, volume free charge density ρ^E, mass density ρ^0 and the second Piola-Kirchhoff stress $T^S{}_{KL}$ [15]. It raises by the transformation of symmetric tensor $\tau^S{}_{mk}$ in the current state to reference state, and relate the traction force with areas, both in the reference state. While the first Piola-Kirchhoff stress is connecting the traction force and electric force in the current state with regions in the reference state.

This section will describe the **transformation of energy conservation from the current state to the reference state**, using Eq. (8), the symmetric tensor modifies Eq. (38).

$$\rho\dot{\chi} = \tau^S{}_{mk}u_{k,m} - P_m E_k u_{k,m} - \dot{E}_m P_m \tag{39}$$

2.2.1 Electric field and gradient of potential

To transform the electric field to a reference state, here will use follow:

$$W_K = E_m y_{m,K} \tag{40}$$

The gradient of the potential W_K is multiplying both sides by $X_{K,m}$

$$W_K X_{K,m} = E_m y_{m,K} X_{K,m} = E_{\mathrm{m}} \frac{\partial y_m}{\partial X_K}\frac{\partial X_K}{\partial y_m} \tag{41}$$

Therefore,

$$E_m = W_K X_{K,m} \tag{42}$$

The derivative respect to time of E_m becomes

$$\dot{E}_m = \frac{d}{dt}\left[W_K X_{K,m}\right] = \frac{d}{dt}(W_K)X_{K,m} + W_K\frac{d}{dt}(X_{K,m}) \tag{43}$$

The term $X_{K,m}$ from Eq. (43) is developed as follow

$$X_{K,m} = \delta_{KL} X_{L,m} = \frac{\partial y_k}{\partial X_L} \frac{\partial X_K}{\partial y_k} X_{L,m} \tag{44}$$

Derivative $X_{K,m}$ respect to time

$$\begin{gathered} \frac{d}{dt}(X_{K,m}) = \frac{d}{dt}\left(\frac{\partial y_k}{\partial X_L}\frac{\partial X_K}{\partial y_k} X_{L,m}\right) = \frac{d}{dt}\left(y_{k,L} X_{K,k} X_{L,m}\right) \\ \frac{d}{dt}(X_{K,m}) = \frac{d}{dt}\left(y_{k,L}\right) X_{K,k} X_{L,m} + \frac{d}{dt}(X_{K,k}) y_{k,L} X_{L,m} + \frac{d}{dt}(X_{L,m}) y_{k,L} X_{K,k} \end{gathered} \tag{45}$$

Partial derivate of y and X are written in Leibniz notation.

$$\frac{d}{dt}(X_{K,m}) = u_{k,L} X_{K,k} X_{L,m} + \frac{d}{dt}(X_{K,k}) \frac{\partial y_k}{\partial X_L}\frac{\partial X_L}{\partial y_m} + \frac{d}{dt}(X_{L,m}) \frac{\partial y_k}{\partial X_L}\frac{\partial X_K}{\partial y_k} \tag{46}$$

The products of partial derivate are reduced to Kronecker delta.

$$\frac{d}{dt}(X_{K,m}) = u_{k,L} X_{K,k} X_{L,m} + \frac{d}{dt}(X_{K,k}) \delta_{km} + \frac{d}{dt}(X_{L,m}) \delta_{KL} \tag{47}$$

The index into $X_{K,k}$ and $X_{L,m}$ were exchanging due to commutation Kronecker deltas.

$$\frac{d}{dt}(X_{K,m}) = u_{k,L} X_{K,k} X_{L,m} + \frac{d}{dt}(X_{K,m}) + \frac{d}{dt}(X_{K,m}) \tag{48}$$

In Eq. (48) was delete the term $\frac{d}{dt}(X_{K,m})$ in both sides

$$0 = u_{k,L} X_{K,k} X_{L,m} + \frac{d}{dt}(X_{K,m}) \tag{49}$$

Clearing $\frac{d}{dt}(X_{K,m})$ we obtain

$$\frac{d}{dt}(X_{K,m}) = -u_{k,L} X_{K,k} X_{L,m} \tag{50}$$

Substituting Eq. (50) into Eq. (43) becomes

$$\dot{E}_m = \frac{d}{dt}[W_K X_{K,m}] = \dot{W}_K X_{K,m} - u_{k,L} X_{K,k} X_{L,m} W_K \tag{51}$$

Then, we replace the Eq. (51) into Eq. (39).

$$\rho\dot{\chi} = \tau^S{}_{mk} u_{k,m} - P_m E_k u_{k,m} - P_m\left(\dot{W}_K X_{K,m} - u_{k,L} X_{K,k} X_{L,m} W_K\right) \tag{52}$$

The index L was changed by k, into $u_{k,L}$, due to $X_{L,m}$

$$\rho\dot{\chi} = \tau^S{}_{mk} u_{k,m} - P_m E_k u_{k,m} - P_m\left(\dot{W}_K X_{K,m} - u_{k,m} X_{K,k} W_K\right) \tag{53}$$

In Eq. (53), the term $X_{K,k} W_K$ is the electric field concerning the current state.

$$\rho\dot{\chi} = \tau^S{}_{mk} u_{k,m} - P_m E_k u_{k,m} - P_m \dot{W}_K X_{K,m} + P_m u_{k,m} E_k \tag{54}$$

From Eq. (54) the term $P_m E_k u_{k,m}$ was removes to get

$$\rho\dot{\chi} = \tau^S{}_{mk} u_{k,m} - P_m \dot{W}_K X_{K,m} \tag{55}$$

2.2.2 Polarization vector

In this subsection, we will perform the transformation of the polarization vector to the reference state.

$$P_L = J X_{L,i} P_i \tag{56}$$

Where J is the Jacobian, multiplying Eq. (56) by $J^{-1} y_{m,L}$ we obtain

$$J^{-1} y_{m,L} P_L = J^{-1} y_{m,L} J X_{L,i} P_i = \delta_{mi} P_i \tag{57}$$

To get

$$P_m = J^{-1} y_{m,L} P_L \tag{58}$$

From Eq. (58) into Eq. (55) results in

$$\begin{aligned} \rho\dot{\chi} &= \tau^S{}_{mk} u_{k,m} - J^{-1} y_{m,L} P_L \dot{W}_K X_{K,m} \\ \rho\dot{\chi} &= \tau^S{}_{mk} u_{k,m} - J^{-1} P_L \dot{W}_K \frac{\partial y_m}{\partial X_L} \frac{\partial X_K}{\partial y_m} \\ \rho\dot{\chi} &= \tau^S{}_{mk} u_{k,m} - J^{-1} P_L \dot{W}_K \delta_{KL} \\ \rho\dot{\chi} &= \tau^S{}_{mk} u_{k,m} - J^{-1} P_K \dot{W}_K \end{aligned} \tag{59}$$

Until now, in Eq. (59), we have obtained a partial transformation, and still missing transform the symmetric Cauchy stress tensor $\tau^S{}_{mk}$.

2.2.3 Second Piola-Kirchhoff stress

The symmetric tensor $\tau^S{}_{mk}$ is related with second Piola-Kirchhoff stress $T^S{}_{KL}$ through a reverse transformation as follow:

$$\tau^S{}_{mk} = J^{-1} y_{m,K} y_{k,L} T^S{}_{KL} \tag{60}$$

Eq. (60) into Eq. (59) results in

$$\rho\dot{\chi} = J^{-1} y_{m,K} y_{k,L} T^S{}_{KL} u_{k,m} - J^{-1} P_K \dot{W}_K \tag{61}$$

The gradient of velocity $u_{k,m}$ can be separated on antisymmetric tensor $\omega_{mk} = \frac{1}{2}(u_{k,m} - u_{m,k})$ plus a symmetric tensor $d_{mk} = \frac{1}{2}(u_{k,m} + u_{m,k})$.

$$\begin{aligned} \rho\dot{\chi} &= J^{-1} y_{m,K} y_{k,L} T^S{}_{KL} (\omega_{mk} + d_{mk}) - J^{-1} P_K \dot{W}_K \\ \rho\dot{\chi} &= J^{-1} y_{m,K} y_{k,L} T^S{}_{KL} \omega_{mk} + J^{-1} y_{m,K} y_{k,L} T^S{}_{KL} d_{mk} - J^{-1} P_K \dot{W}_K \end{aligned} \tag{62}$$

From Eq. (61), the product between symmetric tensor $T^S{}_{KL}$ and antisymmetric tensor ω_{mk} result be null

$$\rho\dot{\chi} = J^{-1}y_{m,K}y_{k,L}T^{S}{}_{KL}d_{mk} - J^{-1}P_{K}\dot{W}_{K} \tag{63}$$

The term $y_{m,K}y_{k,L}d_{mk}$ will be solved as following

$$y_{m,K}y_{k,L}d_{mk} = y_{m,K}y_{k,L}\frac{1}{2}(u_{k,m} + u_{m,k}) = \frac{1}{2}\left(u_{k,m}y_{m,K}y_{k,L} + u_{m,k}y_{m,K}y_{k,L}\right)$$

$$y_{m,K}y_{k,L}d_{mk} = \frac{1}{2}\left(\frac{\partial u_k}{\partial y_m}\frac{\partial y_m}{\partial X_K}\frac{\partial y_k}{\partial X_L} + \frac{\partial u_m}{\partial y_k}\frac{\partial y_m}{\partial X_K}\frac{\partial y_k}{\partial X_L}\right) = \frac{1}{2}\left(\frac{\partial u_k}{\partial X_K}\frac{\partial y_k}{\partial X_L} + \frac{\partial u_m}{\partial X_L}\frac{\partial y_m}{\partial X_K}\right)$$

$$y_{m,K}y_{k,L}d_{mk} = \frac{1}{2}\left(u_{k,K}y_{k,L} + u_{m,L}y_{m,K}\right) \tag{64}$$

We interchange the index k to m in $u_{k,K}$.

$$\begin{aligned} y_{m,K}y_{k,L}d_{mk} &= \frac{1}{2}\left(u_{m,K}y_{k,L} + u_{m,L}y_{m,K}\right) = \frac{1}{2}\left(\dot{y}_{m,K}y_{k,L} + \dot{y}_{m,L}y_{\mathrm{m},K}\right) \\ y_{m,K}y_{k,L}d_{mk} &= \frac{1}{2}\frac{d}{dt}\left(y_{m,K}y_{k,L}\right) = \frac{d}{dt}\left[\frac{1}{2}\left(y_{m,K}y_{k,L} - \delta_{KL}\right)\right] \end{aligned} \tag{65}$$

With $m = k$ the term $\frac{1}{2}\left(y_{m,K}y_{k,L} - \delta_{KL}\right)$ is known as the finite strain tensor E_{KL} in the reference state, and Ȩ whit an uppercase index will represent the electric field vector in the reference state. Then, we reduce $y_{m,K}y_{k,L}d_{mk}$ to:

$$y_{m,K}y_{k,L}d_{mk} = \frac{d}{dt}(E_{KL}) = \dot{E}_{KL} \tag{66}$$

Substituting Eq. (66) into Eq. (63), we obtain

$$\rho\dot{\chi} = J^{-1}T^{S}{}_{KL}\dot{E}_{KL} - J^{-1}P_{K}\dot{W}_{K} \tag{67}$$

Factoring the inverse of Jacobian, we get

$$\rho\dot{\chi} = J^{-1}(T^{S}{}_{KL}\dot{E}_{KL} - P_{K}\dot{W}_{K}) \tag{68}$$

Multiplying both sides into Eq. (68) by the Jacobian gives

$$J\rho\dot{\chi} = JJ^{-1}(T^{S}{}_{KL}\dot{E}_{KL} - P_{K}\dot{W}_{K}) \tag{69}$$

Using mass transformation to the reference state $\rho^0 = \rho J$ into Eq. (69), we get a new equation for energy conservation in terms of physical quantities in the reference state. Symmetric tensor $T^{S}{}_{KL}$, strain tensor E_{KL}, polarization P_K and gradient of potential W_K.

$$\rho^0\dot{\chi} = T^{S}{}_{KL}\dot{E}_{KL} - P_{K}\dot{W}_{K} \tag{70}$$

2.3 Constitutional equations from free energy

The conservation laws are valid for any piezoelectric material, including cement-based composites. However, a specific material's piezoelectric properties are determined by a set of functions that describes free energy, symmetric tensor, and polarization. Once we replace these functions into Eq. (70), we will get the piezoelectricity's constitutional equations. Take into account Eq. (70), we can propose the next dependence to the functions

$$\begin{aligned} \chi &= \chi(E_{KL}, W_K) \\ T^S{}_{KL} &= T^S{}_{KL}(E_{KL}, W_K) \\ P_K &= P_K(E_{KL}, W_K) \end{aligned} \tag{71}$$

Derivation respect to time the free energy into Eq. (71) as follow

$$\dot{\chi} = \frac{\partial \chi}{\partial E_{KL}} \dot{E}_{KL} + \frac{\partial \chi}{\partial W_K} \dot{W}_K \tag{72}$$

Substituting Eq. (72) into Eq. (70), we obtain

$$\rho^0 \frac{\partial \chi}{\partial E_{KL}} \dot{E}_{KL} + \rho^0 \frac{\partial \chi}{\partial W_K} \dot{W}_K = T^S{}_{KL} \dot{E}_{KL} - P_K \dot{W}_K \tag{73}$$

Both sides of Eq. (73) were compared to deduce two transformations, which resulting symmetric tensor $T^S{}_{KL}$ and polarization P_K. The transformations use free energy as a generating function, as shown in Eq. (75).

$$T^S{}_{KL} = \rho^0 \frac{\partial \chi}{\partial E_{KL}} \tag{74}$$

$$P_K = -\rho^0 \frac{\partial \chi}{\partial W_K} \tag{75}$$

The mathematical structure of the free energy function will define the order of constitutional equations. There are functions for the free energy of piezoelectric materials from order 1 to order 3 [15]. It means that piezoelectric material behavior depends on the free energy function and its parameters. Here is an example of free energy function with order three

$$\begin{aligned} \rho^0 \chi = {} & \frac{1}{2} c_{ABCD} E_{AB} E_{CD} - e_{ABC} W_A E_{BC} - \frac{1}{2} \chi^E{}_{AB} W_A W_B + \frac{1}{6} c_{ABCDEF} E_{AB} E_{CD} E_{EF} \\ & + \frac{1}{2} d_{ABCDE} W_A E_{BC} E_{DE} - \frac{1}{2} b_{ABCD} W_A W_B E_{CD} - \frac{1}{6} \chi^E{}_{ABC} W_A W_B W_C \\ & + \frac{1}{24} c_{ABCDEFGH} E_{AB} E_{CD} E_{EF} E_{GH} + \frac{1}{6} d_{ABCDEFG} W_A E_{BC} E_{DE} E_{FG} \\ & + \frac{1}{4} a_{ABCDEF} W_A W_B E_{CD} E_{EF} + \frac{1}{6} d_{ABCDE} W_A W_B W_C E_{DE} \\ & - \frac{1}{24} \chi^E{}_{ABCD} W_A W_B W_C W_D + \dots, \end{aligned} \tag{76}$$

The parameters are called elasticity c, piezoelectric e, electric permeability χ^E, odd electrolytic d, electrostrictive b, and electroelastic force even a.

2.3.1 The linear approach of piezoelectricity

We take on order one approach from Eq. (76) to free energy χ. Then, replacing it in Eq. (74) and Eq. (75) to obtain

$$T^S{}_{AB} = \frac{\partial}{\partial E_{AB}}\left(\frac{1}{2}c_{ABCD}E_{AB}E_{CD} - e_{ABC}W_A E_{BC} - \frac{1}{2}\chi^E{}_{AB}W_A W_B\right) \tag{77}$$

$$P_A = -\frac{\partial}{\partial W_A}\left(\frac{1}{2}c_{ABCD}E_{AB}E_{CD} - e_{ABC}W_A E_{BC} - \frac{1}{2}\chi^E{}_{AB}W_A W_B\right) \tag{78}$$

The approximation is possible if we consider an infinitesimal deformation, weak electric field, and low amplitude displacements around the reference state. Hence, it approaches require a nomenclature exchange for physical quantities. Thus, second Piola-Kirchhoff stress will be replaced by infinitesimal Cauchy stress tensor $T^S{}_{KL} \rightarrow T_{ij}$; finite strain tensor will be exchanged by infinitesimal strain tensor $E_{KL} \rightarrow S_{kl}$; potential gradient, polarization, and displacement electric vector are similar either reference or current state: $W_K \rightarrow E_k$, $P_L \rightarrow P_i$, and $\mathcal{D}_L \rightarrow D_i$. Then, Eqs. (77) and (78) follow:

$$T_{ij} = \frac{\partial}{\partial S_{ij}}\left(\frac{1}{2}c_{ijkl}S_{ij}S_{kl} - e_{ijk}E_i S_{jk} - \frac{1}{2}\chi^E{}_{ij}E_i E_j\right) \tag{79}$$

$$P_i = -\frac{\partial}{\partial E_i}\left(\frac{1}{2}c_{ijkl}S_{ij}S_{kl} - e_{ijk}E_i S_{jk} - \frac{1}{2}\chi^E{}_{ij}E_i E_j\right) \tag{80}$$

Here is considering symmetry to parameters elastic c_{ijkl}, piezoelectric e_{kij}, and electric χ_{ik} when they have odd permutations. Differentiating the Eq. (79) and Eq. (80), we obtain

$$T_{ij} = c_{ijkl}S_{kl} - e_{kij}E_k \tag{81}$$

$$P_i = e_{ikl}S_{kl} + \chi^E{}_{ik}E_k \tag{82}$$

The polarization can be written in terms of electric displacement vector too.

$$P_i = D_i - \varepsilon_0 E_i \tag{83}$$

From Eq. (83) into Eq. (82) gives

$$D_i - \varepsilon_0 E_i = e_{ikl}S_{kl} + \chi^E{}_{ik}E_k \tag{84}$$

Solving D_i,

$$D_i = e_{ikl}S_{kl} + \varepsilon_0 E_i + \chi^E{}_{ik}E_k = e_{ikl}S_{kl} + \varepsilon_0\delta_{ik}E_i + \chi^E{}_{ik}E_k \tag{85}$$

Factoring E_k,

$$D_i = e_{ikl}S_{kl} + \left(\varepsilon_0\delta_{ik} + \chi^E{}_{ik}\right)E_k \tag{86}$$

where the term $\varepsilon_0\delta_{ik} + \chi^E{}_{ik}$ is defined as dielectric constant ε_{ik}. Finally, we have the linear constitutional equation for the electric displacement vector.

$$D_i = e_{ikl}S_{kl} + \varepsilon_{ik}E_k \tag{87}$$

We have seen several forms to present the linear constitutional equations in piezoelectricity. Next, we include another form of constitutional equations shown in the IEEE standard for piezoelectricity. It can be obtained inverting the matrix formed by Eq. (81) and Eq. (82).

$$D_i = d_{ikl}S_{kl} + \varepsilon^T_{ik}E_k \tag{88}$$

$$S_{ij} = s^D_{ijkl}T_{kl} + g_{kij}D_k \tag{89}$$

The electromechanical properties are defined by piezoelectric charge d_{ikl} and voltage g_{kij} constants. Unlike parameters c_{ijkl} and e_{ikl} These new piezoelectric constants are taken out directly from experiments, as shown in the next section.

2.4 Electromechanical and electrical properties of cement-based composites

Incorporating piezoelectric nanocomposites into cement paste improves its piezoelectric and mechanical properties [16] due to increased deformable crystal structures. Zeolites, oxides, and carbon nanotubes are the most used cement-based composites to improve these properties [17]. Chen et al. also report some piezoelectric parameters of cement-based composites such as piezoelectric charge d_{33}, voltage g_{33}. And the coupling factor K_t. As was mentioned in the previous section, these piezoelectric parameters come from linear piezoelectricity theory. However, the crystalline structure of Calcium Silicate Hydrate (C-S-H) that compose the cement is a complex system described by linear theory. It could also be combined with statistical physics and mean-field homogenization theory tools to get the macroscale properties [18]. Here are show piezoelectric and electrical parameters of gold nanoparticles mixed to cement paste, which we hope to lead to our system's constitutional equations.

Next, we introduce a brief description of the gold nanoparticles' physical synthesis [19, 20]. They are produced by laser ablation at 532 nm. A gold plate at 99.9999% purity is put inside a beaker filled with 50 mL of ultrapure water. Then, the pulse laser spot with an energy of 30 mJ beats the gold plate by 10 minutes, as shown in **Figure 3**.

At the time, the gold nanoparticles were brought to be characterized by dynamical light scattering (DLS). If not done quickly, the gold nanoparticles were agglomerated. These measures are required because the gold nanoparticles directly affect the piezoelectric properties of cement cylinders. Some results of gold nanoparticle sizes are shown in **Figure 4**.

Also, the gold nanoparticles in water must be mixed quickly with the cement. The ratio of water/cement used was 0.47 mL/g. Then, the admixture was poured into cylindrical molds that contained copper wires as follows in **Figure 5**.

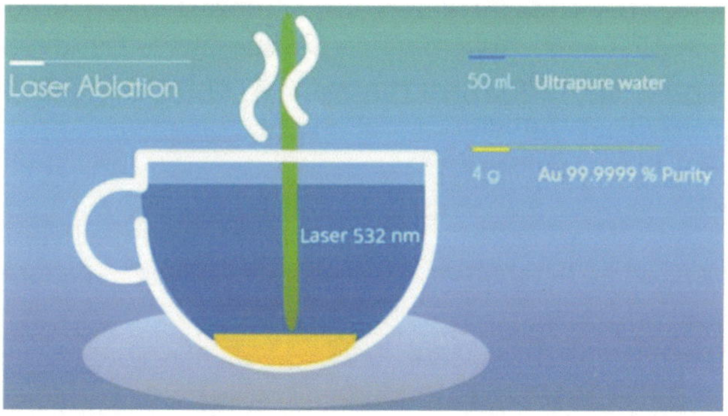

Figure 3.
Scheme of nanoparticle physical synthesis by laser ablation.

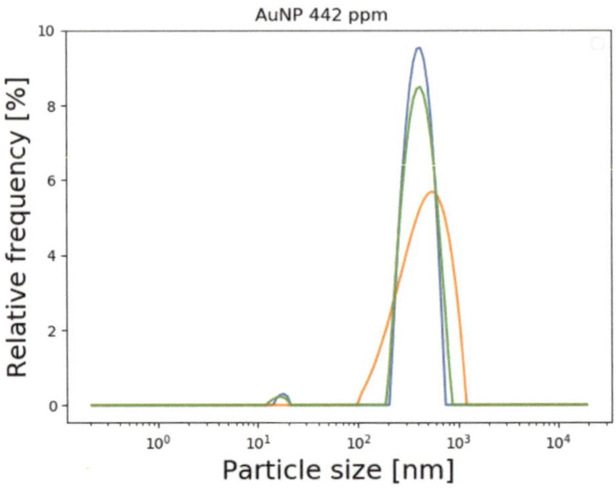

Figure 4.
The particle size distribution of gold nanoparticles suspended in water to concentration 442 ppm.

Figure 5.
Molds and dimensions of cement cylinders.

The cement cylinders were dried one day. Then it leaves curing for 28 days and finally to thermal treatment one day more. After 14 days, electromechanical measurements were performed, as shown in **Figure 6**.

Electromechanical measurements consist of two measurements performed in parallel: the cement cylinders under compressive strength test in the axial direction, open circuit potential (OCP) measurements in the electrodes of cement cylinders. From mechanical and electrical data, we calculated an electroelastic parameter with units $[mV/kN]$, it has the same interpretation of piezoelectric parameter e in linear theory. From **Figure 7**, an example of voltage-force curves for identically cement samples with gold nanoparticles is shown. We did get from the above measurements the axial elasticity parameter:

$$Y = 323.5 \pm 75.3 \left[kN/m^2\right] \tag{90}$$

The axial piezoelectric parameter:

$$\gamma = -20.5 \pm 6.9 \, [mV/kN]. \tag{91}$$

For a total deformation $S = 0.57 \pm 0.09 \, [mm]$ in the axial direction.

The electrical properties of cement cylinders were obtained from the imaginary part of impedance; an example of these curves in **Figure 8**. From impedance data

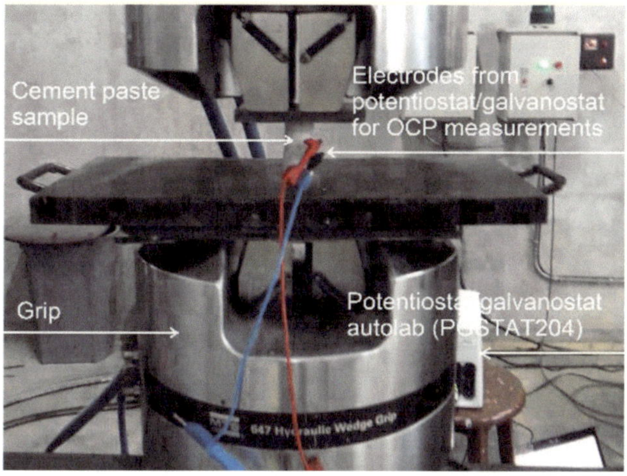

Figure 6.
Experimental setup of electromechanical measurements.

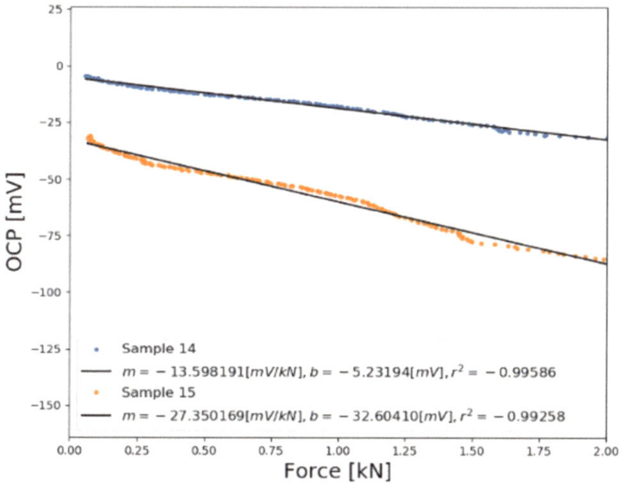

Figure 7.
OCP-force curves from cement cylinders with gold nanoparticles concentrated to 658 ppm.

can perform a transformation to get a real part of the capacitance C'. It has frequency dependence as follow

$$C'(\omega) = \frac{1}{\omega Z''} \tag{92}$$

The geometry of copper electrodes (an approximation to parallel plates) is related to capacitance. Therefore, we can calculate the dielectric parameter ε since 1 MHz; this parameter is a real number that depends on the frequency and is given by

$$\varepsilon(\omega)\varepsilon_0 = \frac{d * C'}{A} \tag{93}$$

where ε_0 is the electric permittivity of free space, A is the transversal section, and d is the thickness between electrodes.

From the data in **Figure 8** and Eq. (92) and Eq. (93), we obtain the dielectric constant:

$$\varepsilon = (939.6 \pm 82.9)\varepsilon_0 \tag{94}$$

Where ε_0 has unit $[F/m]$. The piezoelectric and electrical properties of cement paste mixed with gold nanoparticles exhibit reproducibility and linearity of the piezoelectric parameter.

2.5 Future studies and remarks

The Piezoelectric parameters are an initial point to beginning a new connection with piezoelectricity theory by inverse modeling and constructing new free energy functions and constitutional equations. To catch out with researchers in this scope, we suggest thinking about the next research questions; how is the piezoelectric parameter presented related to the piezoelectric parameter formulated by linear theory for piezoelectricity? Is the free energy function of order one sufficient to describe cement paste's piezoelectric with gold nanoparticles? How to develop a new function for free energy that models cement paste's piezoelectric behavior of cement paste with gold nanoparticles?

In this chapter, we have intended to contribute to the theory of piezoelectricity for large deformations without including an energy function. **Figure 9** shows a possible use around IoT as intelligent sensing of devices based on cement-based composites' piezoresistivity. Without reaching into depth in the technical and engineering aspect that smart construction, active sensing system entails; we highlight how the Eqs. (88) and (89) that relate the electromechanical properties and that are defined by piezoelectric charge d_{ikl} and voltage g_{kij} constants are present as indicators to improve the detection resolution in large structures with large deformations.

The sensors analyze the deformations, temperature, relative humidity, and other critical parameters of the concrete in real-time. This data is captured via wireless communication (WAN/BLE) and deployed on a secure and scalable platform (Cloud) capable of collecting data to facilitate remote decision making with

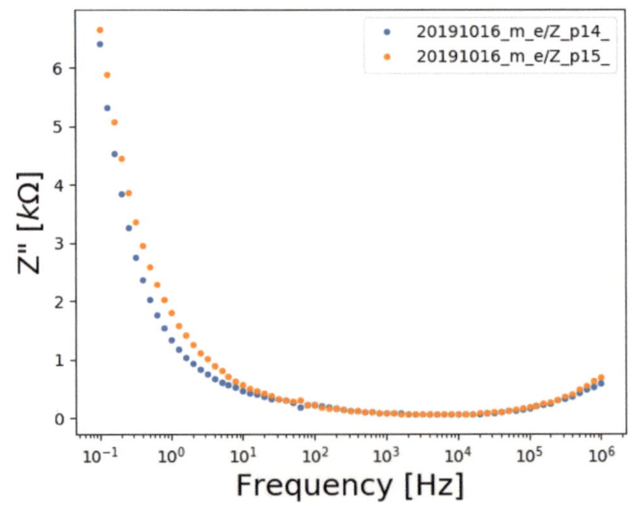

Figure 8.
The imaginary part of electrical impedance represented in a Bode plot was performed on two cement cylinders with gold nanoparticles concentrated to 658 ppm.

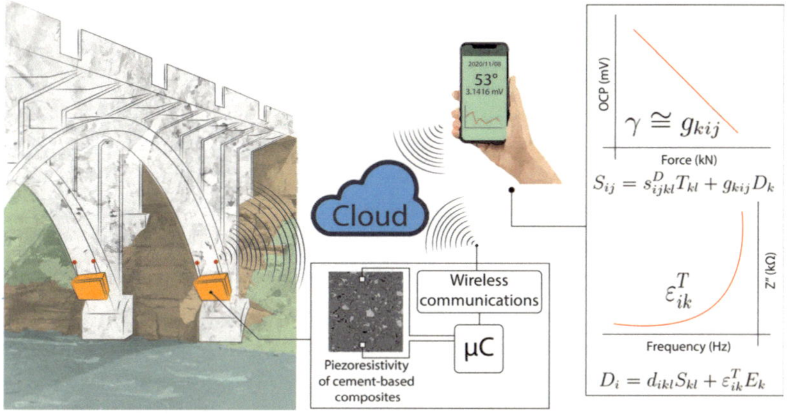

Figure 9.
The image shows a network of IoT sensors based on cement-based composites piezoresistivity as an active part of smart construction.

information from deep within the concrete. The experimental control of the NPs embedded within the cement paste's dispersions and piezoresistive responses is essential to have a good signal-to-noise ratio within the sensing. Knowing the coupling between the electromechanical equations from a theoretical approach is another crucial factor in making viable these technological solutions.

3. Conclusions

This chapter proposed a mathematical physicist construction of the linear theory of piezoelectricity since classical movement laws and the conservation of their physical quantities (mass, charge, linear momentum, angular momentum, and energy) over time. This construction takes parts of Eringen, Tiersten, and Yang's research without including the variational formulation or energy functional to deduce the constitutional equations. We have also presented some results of piezoelectric and dielectric constants obtained for cement mixed to gold nanoparticles. We got the axial elasticity parameter $Y = 323.5 \pm 75.3\ [kN/m^2]$, the electroelastic parameter $\gamma = -20.5 \pm 6.9\ \ [\text{mV/kN}]$, and dielectric constant $\varepsilon = (939.6 \pm 82.9)\varepsilon_0[F/m]$, which can be compared with parameters s^D_{ijkl}, g_{kij} and ε^T_{ik} respectively presents into constitutional equations discussed in the chapter.

Acknowledgements

We would like to thank the Vice-rector for research in project N 2676 of the Universidad Industrial de Santander, the CIMBIOS research group for the ablation laser system (Universidad Industrial de Santander), and the CA Perez-Lopez for his support in the editing of images of the Department of Electrical and Electronic Engineering of the Universidad de los Andes Colombia.

Conflict of interest

The authors declared no potential conflicts of interest concerning the research, authorship, and/or publication of this book chapter.

Appendices and nomenclature

In the reference state, the continuum has a volume V, and mass density ρ^0.

In the current state, the continuum has a volume v, mass density ρ, electronic charge density μ^e and lattice charge density μ^l. Besides, In the current state with infinitesimal displacement η, the electronic charge does not change its volume.

The capital letter in the index is for the reference state X_K And the lowercase letters to the current state y_i. Also, the index in the physics quantities can denote a vector. For example X_K, y_i, u_i; or a tensor, for example E_{KL}, τ_{jk}. Another form to present a vector quantity is the right-pointing arrow $\vec{y}$.

The velocity of the continuum is denoted by lower case letter u, and just makes sense in the current state.

The partial derivate is denoted by comma separation in the indexes.
For example $y_{i,i}$.

Author details

Daniel A. Triana-Camacho[1], Jorge H. Quintero-Orozco[1] and Jaime A. Perez-Taborda[2]*

1 CIMBIOS Group, Physics Department, Universidad Industrial de Santander, Bucaramanga, Colombia

2 Department of Electrical and Electronic Engineering, Universidad de los Andes, Bogotá, Colombia

*Address all correspondence to: jaimeandres@protonmail.com

References

[1] Katzir, S. The beginnings of piezoelectricity: A study in mundane physics. Springer Science & Business Media; 2007. 273 p. DOI: 10.1007/978-1-4020-4670-4

[2] Voigt, W. Piezo-und pyroelectricität, dielectrische influenz und electrostriction bei krystallen ohne symmetriecentrum. Annalen Der Physik. 1985;291(8):701-731. DOI: 10.1002/andp.18952910812

[3] Eringen, A. C. On the foundations of electroelastostatics. 1963;1(1):127-153. DOI: 10.1016/0020-7225(63)90028-4

[4] Toupin, R. A. The elastic dielectric. Journal of Rational Mechanics and Analysis. 1956;5(6):849-915. Retrieved from http://www.jstor.org/stable/24900192

[5] Tiersten, H. F. On the non-linear equations of thermo-electroelasticity. International Journal of Engineering Science. 1972;9(7):587-604. DOI: 10.1016/0020-7225(71)90062-0

[6] Martin, R. M. Piezoelectricity. Physical Review B. 1972;5(4):1607-1613. DOI: 10.1103/PhysRevB.5.1607

[7] Casamento, J., Chang, C. S., Shao, Y. T., Wright, J., Muller, D. A., Xing, H., & Jena, D. Structural and piezoelectric properties of ultra-thin ScxAl1− xN films grown on GaN by molecular beam epitaxy. Applied Physics Letters. 2020; 117(11):112101. DOI: 10.1063/5.0013943

[8] Auld, B. A. Acoustic fields and waves in solids Рипол Классик. John Wiley & Sons, New York. 1973.

[9] Ma, J., Ren, J., Jia, Y., Wu, Z., Chen, L., Haugen, N. O., ... & Liu, Y. High efficiency bi-harvesting light/vibration energy using piezoelectric zinc oxide nanorods for dye decomposition. Nano Energy. 2019;62:376-383. DOI: 10.1016/j.nanoen.2019.05.058

[10] Bechmann, R. Elastic and piezoelectric constants of alpha-quartz. Physical Review, 1958;110(1): 1060-1061. DOI: 10.1103/PhysRev.110.1060

[11] Tiersten, H. F. Linear piezoelectric plate vibrations: Elements of the linear theory of piezoelectricity and the vibrations piezoelectric plates. Springer, Boston, MA. 2013. DOI: 10.1007/978-1-4899-6453-3

[12] Yang, J. On the derivation of electric body force, couple and power in an electroelastic body. Acta Mechanica Solida Sinica. 2015;28(6):613-617.

[13] Yang, J. Differential derivation of momentum and energy equations in electroelasticity. Acta Mechanica Solida Sinica. 2017;30(1), 21-26.

[14] Abeyaratne, R., & Knowles, J. K. A continuum model of a thermoelastic solid capable of undergoing phase transitions. Journal of the Mechanics and Physics of Solids. 1993;41(3): 541-571. DOI: 10.1016/0022-5096(93)90048-K

[15] Tiersten, H. F. Non-linear electroelastic equations cubic in the small field variables. The Journal of the Acoustical Society of America. 1975;57(3):660-666. DOI:10.1121/1.380490

[16] Chen, J., Qiu, Q., Han, Y., & Lau, D. Piezoelectric materials for sustainable building structures: Fundamentals and applications. Renewable and Sustainable Energy Reviews. 2019;101:14-25. DOI: 10.1016/j.rser.2018.09.038

[17] Paul, S. C., van Rooyen, A. S., van Zijl, Gideon P. A. G., & Petrik, L. F. Properties of cement-based composites using nanoparticles: A comprehensive review. Construction and Building Materials. 2018;189:1019-1034. DOI: 10.1016/j.conbuildmat.2018.09.062

[18] Qomi, M. J. A., Ulm, F. J., & Pellenq, R. J. M. Physical origins of thermal properties of cement paste. Physical Review Applied. 2015;3(6): 064010. DOI: 10.1103/ PhysRevApplied.3.064010

[19] Yan, Z., & Chrisey, D. B. Pulsed laser ablation in liquid for micro-/ nanostructure generation. Journal of Photochemistry and Photobiology C: Photochemistry Reviews. 2012;13(3): 204-223. DOI:10.1016/ j.jphotochemrev.2012.04.004

[20] Huang, H., Lai, J., Lu, J., & Li, Z. Pulsed laser ablation of bulk target and particle products in liquid for nanomaterial fabrication. AIP Advances. 2019;9(1):015307. DOI: 10.1063/ 1.5082695

Section 4

Sustainable Applications of Cement

Chapter 10

Sustainable Recycling of Marble Dust as Cement Replacement in Concrete: Advances and Recent Trends

Ahed Habib and Maan Habib

Abstract

In recent years, many researchers in the construction industry had taken up the challenge to incorporate non-biodegradable wastes as partial replacement of cement and/or natural aggregates in the daily production of cement-based materials. Various efforts were intended to understand the influence of using marble dust in concrete due to its availability and a relatively high volume of the generation that causes serious environmental problems. Previous studies have utilized marble dust as a replacement of cement, fine aggregate, or total paste in the concrete and mortar mixtures. In general, several investigations have shown that up to a certain cement replacement ratio, marble dust can positively impact on the strength and microstructure properties of concrete. Furthermore, the results have indicated that the considerably high degree of fineness in the marble dust provides sufficient cohesiveness of mortar and concrete even in low w/c ratio conditions. Hence, this powder can be utilized as a filler to improve the flowability of cement-based materials. Consequently, this chapter aims to summarize recent investigations on the properties of concrete incorporating marble waste as cement replacement materials, highlight the potential gaps in the literature, and propose a prediction model for estimating the compressive and flexural strengths of concrete with marble dust using regression analysis.

Keywords: sustainable materials, marble dust, cement-based materials, cement replacement, mechanical properties

1. Introduction

Throughout the last few decades, considerable efforts in the scientific community were focused on providing sustainable solutions for minimizing non-biodegradable wastes by suggesting innovative waste management plans. Recently, the construction industry has started taking an active role in recycling these materials by utilizing them as a partial replacement of the constituents in cement-based productions, aiming to come up with a green alternative for conventional construction materials.

Stone marble industrial activities, including mining, processing, and finishing, have contributed to the development of several major environmental risks [1].

One of these risks is the disposal of marble wastes that are raised during the production of marble slabs. Currently, the availability and the reasonably high volume of a generation of marble dust has attracted several researchers to conduct investigations on the possibility of utilizing this waste material as a partial replacement of cement [2–4], fine aggregates [5, 6], or total paste in concrete, mortar, and asphaltic mixtures [7, 8]. The main benefit of replacing cement by marble dust comes from the reduction in the cost of the mixture, and CO2 emission related to the production of cement [9].

Previous studies have illustrated that incorporating this material in concrete affects its fresh, mechanical, durability, and porosity properties [10–12]. Its main impact on the strength carrying capacity of concrete is generally considered such that when replacing a low percentage of the cement with marble dust, the strength is improved. However, at high replacement ratios, beyond 10% to 15%, the compressive and tensile strengths of the concrete reduce. Another importance of this material, in addition to its sustainable benefits, comes from the high degree of fineness that allows utilizing it as filler in cement-based mixtures.

Therefore, this chapter is intended to offer a brief review of the utilization and mechanical properties of cement-based materials incorporating marble dust with emphasis on concrete mixtures with marble wastes cement replacement. Another aim of the study is to propose estimation models using multiple regression analysis for the compressive and flexural strengths of marble dust concrete and highlight some observations on the influence of marble dust content and properties on the change in the concrete strength capacity.

2. Engineering properties marble dust

Marble is defined scientifically as a metamorphic rock composed of recrystallized calcite ($CaCO_3$) or dolomite ($CaMg(CO_3)_2$), while commercially as any limestone or dolomite processed and taking a polish, **Figure 1** [13]. During cutting and polishing these stones in the marble factories, a product composed of the marble dust mixed with water referred to in the literature as marble waste slurry is generated. Usually, marble dust is obtained by chemically processing the marble waste slurry to separate the wastewater from the marble dust.

The X-ray powder diffraction (XRD) pattern of marble dust is exhibited in **Figure 2**, which shows calcite ($CaCO_3$) as the main mineral component of this material. Also, **Figure 3** presents an example of a scanning electron microscope (SEM) micrographs of marble dust particles.

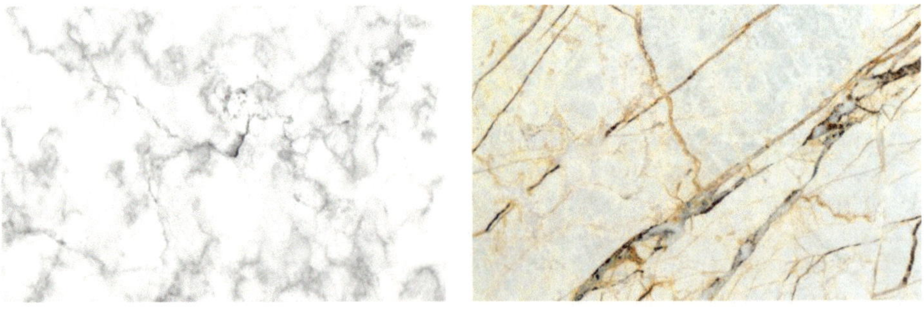

Figure 1.
Typical types of polished marbles.

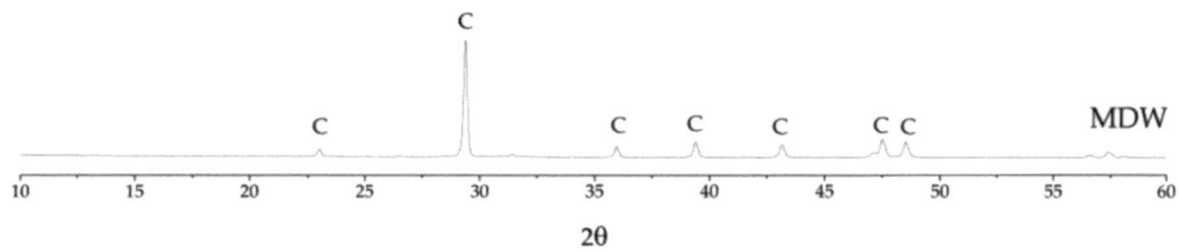

Figure 2.
XRD pattern of marble dust waste as discussed in Julphunthong & Joyklad [14] study.

Figure 3.
SEM image of marble dust as given by Julphunthong & Joyklad [14].

Author/s	Chemical analysis (wt. %)							
	CaO	SiO_2	Fe_2O_3	Al_2O_3	MgO	Na_2O	K_2O	SO_3
[16]	52.45	1.29	0.78	0.39	0.54	-	0.11	-
[17]	40.73	6.01	0.8	0.6	15.21	0.06	0.05	0.09
[18]	41.83	8.38	0.65	0.67	10.36	0.60	0.07	0.33

Table 1.
Chemical compositions of marble dust as reported in the literature.

As reported previously, the specific gravity of this material varies over a wide range between 2.39 and 3.16 due to the difference in the structural and chemical properties of the marble stones [15]. Similarly, the chemical compositions of marble dust various based on the type of stone used. Some examples of the chemical compositions from Gesoğlu et al. [16], Vardhan et al. [17], and Ashish [18] studies can be seen in **Table 1**.

The cutting one-meter cube of marble block into slabs of 2 cm thickness each, 25% of the total amount will turn into fine particles [19]. Previous studies, Gesoğlu et al. [16], Singh et al. [20], and Li et al. [21] presented the particle size distribution of marble wastes as demonstrated in **Figure 4**, in which it depends mainly on the method of cutting the marbles and the size of the produced layer.

3. Types of marble waste utilizations and potential applications

Over the last few years, marble dust has been introduced to various kinds of cement-based materials as a partial replacement of cement [2, 5], fine aggregate [22],

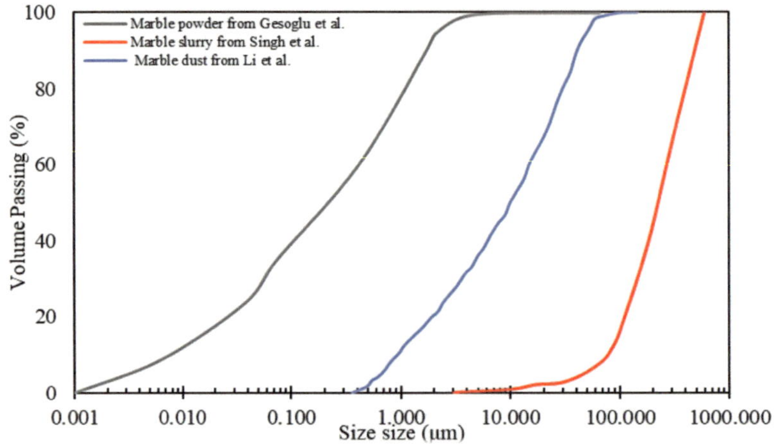

Figure 4.
Particle size distribution of marble wastes used in the literature.

Author/s	Mixture type	Type of replacement
[2–4]	Concrete	Cement
[1]	Concrete	Cement and fine aggregate
[23–25]	Self-compacting concrete	Cement
[26]	High-performance concrete	Cement
[5]	Concrete paving blocks	Fine aggregate
[27]	Concrete	Fine aggregate
[28]	Mortar	Cement
[6]	Mortar	Fine aggregate
[21]	Mortar	Total paste
[29]	Cement composites	Cement

Table 2.
A brief summary of the utilization of marble wastes in the literature.

or total paste [21]. The essential types of mixtures in which marble dust has been utilized can be epitomized as concrete, mortar, cement composites. A summary of the types of marble wastes utilization in cement-based materials can be seen in **Table 2**.

Previously Singh et al. [30] have discussed some of the potential applications of marble wastes in the construction industry. Some of these applications are presented as follows:

- It can be used as a filler for roads and embankment materials where water bound macadam can be laid.

- It can be implemented in the manufacturing process of bricks due to the existence of very fine particles in marble slurry.

- It can be utilized as a partial replacement of cement in concrete due to is the capability of being used as a filler to improve the concrete's properties.

- It can be used in the production of hollow blocks and wall tiles in addition to other clay-based products.

- It can be utilized as a substitute of limestone in various construction materials and industrial applications.

4. Mechanical properties of concrete utilizing marble waste

In this section, the mechanical properties of a concrete mixture incorporating marble waste as partial replacement of cement will be introduced. Previous studies showed, **Figure 5**, that the mechanical properties of concrete mixtures are influenced when marble waste is incorporated. Ergün [3] observed that replacing 5% of the cement content by marble powder results in increasing the compressive strength of concrete by almost 12%. Also, he reported that at the same replacement ratio, a 5% increase in the flexural strength of concrete was achieved. In contrast, higher marble dust content indicated a negative effect on the flexural capacity. Moreover, Munir et al. [31] measured a slight increase in the compressive strength of concrete when 10% of its cement was replaced by marble powder. Vardhan et al. [17] reported that utilizing marble powder as a partial substitution of up to 10% of the cement content in concrete does not have a significant negative influence on the compressive strength. A similar observation was mentioned by Rana et al. [32].

Furthermore, Rana et al. [32] reported a slight reduction in the flexural capacity of concrete when marble wastes replaced up to 10% of the cement. In contrast, higher replacement ratios caused a considerable fall in flexural strength. The modulus of elasticity of concrete with marble powder as substitution of cement was investigated by Soliman [33]. In general, they concluded that up to a 5% ratio, the modulus of elasticity is positively impacted, and beyond this value, the parameter starts to drop slightly. Nevertheless, Usysal & Yilmaz [25] clarified that an increase in the modulus of elasticity when up to 20% of the cement was changed by marble powder in a self-compacting concrete mixture, but a slight decline is observed at 30% replacement ratio. Kumar and Kumar [2] measured an increase in the splitting tensile and flexural strengths by 9.21% and 7.5%, respectively, at 15% cement substitution, while the compressive strength was reduced by 9.06%. On the other hand, using a 20% replacement ratio caused a lowering in the compressive and tensile strengths as compared to the control specimens. This reduction in the strength properties of concrete incorporating high content of marble wastes, beyond 10% in most cases, as a substitution of cement, can be attributed to the decrease in the cement content [3].

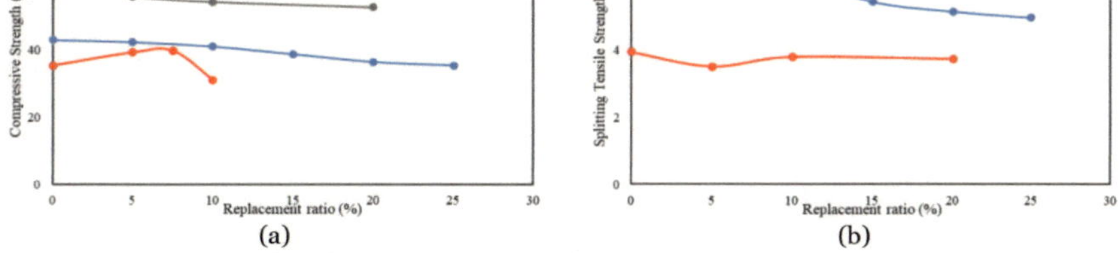

Figure 5.
An example of some mechanical properties of concrete utilizing marble waste as cement replacement (a) compressive strength, (b) splitting tensile strength.

5. Prediction of concrete compressive strength

In this section, a prediction model for the compressive strength of concrete incorporating marble wastes as partial replacement of cement will be addressed.

5.1 Collected data

The collected dataset for generating the prediction models in this study are displayed in **Tables 3** and **4**. First of all, the papers that discussed the utilization of marble dust were collected. Thereafter, those studies that investigated the compressive and flexural strength of concrete utilizing marble waste as cement replacement were shortlisted, and their data were obtained using GetData graph digitizer software [36].

Several parameters are going to be used in developing the estimation model for marble dust compressive strength (f_{cm}) and its flexural strength (f_{tm}). These parameters are the control compressive strength (f_{cc}) or flexural strength (f_{tc}), marble dust content (MD_C), its CaO content (MD_{CaO}), and its specific gravity (MD_{SG}). The last three inputs are basically used to take the effect of marble waste properties on concrete behavior. Because this material does not have a standardized characteristic in which its properties depend on the type of rocks being processed in the factories.

5.2 Multiple linear regression

As mentioned previously, the multiple linear regression method will be used for building the mathematical expression of the estimation model. In general, it is a statistical way to establish a linear relationship between a dependent variable and two or more independent predictors [37]. The mathematical model that describes this method of estimation was discussed by Achen [38], as written in (Eq. (1)).

$$y_i = \beta_0 + \beta_1 x_{1i} + \dots + \beta_k x_{ki} + \varepsilon_i \tag{1}$$

where y_i is the ith observation on the dependent variable; $x_{1i}, \dots, x_{ki}$ are the ith observations on the independent variables; β_0 is an intercept term; $\beta_1, \dots, \beta_k$ are the coefficients to be estimated; and ε_i is a random error component of the ith observation, also known as the residual.

After creating a prediction model, it is essential to test its goodness-of-fit. On this matter, the coefficient of determination (Eq. (2)) was adopted in this study in which values closer to one represents a good fitting model.

$$R^2 = 1 - \frac{\sum (y_i - \hat{y}_i)^2}{\sum (y_i - \bar{y})^2} \tag{2}$$

where y_i is the actual value, $\hat{y}_i$ is the predicted one, and $\bar{y}$ is the mean of the actual values.

Thereafter, the adequacy of the prediction model is evaluated by conducting a residual analysis through first plotting the residuals, (Eq. (3)), against one of the independent variables and then scaling these values using the standardized residuals method (Eq. (4)) to obtain potential outliers.

$$e_i = (y_i - \hat{y}_i) \tag{3}$$

Author/s	Marble waste properties			Compressive strength		
	Content (kg)	CaO (%)	Specific Gravity	f_{cc} (MPa)	f_{cm} (MPa)	Δf_c (%)
Ergün [3]	15	51.7	2.68	35.4	39.4	-10.15
Ergün [3]	22.5	51.7	2.68	35.4	39.9	-11.28
Ergün [3]	30	51.7	2.68	35.4	31.1	13.83
Gesoğlu et al. [16]	26	52.45	2.71	58.57	55.94	4.70
Gesoğlu et al. [16]	52	52.45	2.71	58.57	54.34	7.78
Gesoğlu et al. [16]	104	52.45	2.71	58.57	52.75	11.03
Rana et al. [32]	20.25	65.2	2.87	43	42.4	1.42
Rana et al. [32]	40.5	65.2	2.87	43	41.1	4.62
Rana et al. [32]	60.75	65.2	2.87	43	38.8	10.82
Rana et al. [32]	81	65.2	2.87	43	36.44	18.00
Rana et al. [32]	101.25	65.2	2.87	43	35.4	21.47
Sardinha et al. [34]	15.4	54.2	2.73	39.2	37.3	5.09
Sardinha et al. [34]	30.7	54.2	2.73	39.2	34.3	14.29
Sardinha et al. [34]	61.4	54.2	2.73	39.2	28	40.00
Sardinha et al. [34]	15.4	54.2	2.73	52.1	46.2	12.77
Sardinha et al. [34]	30.7	54.2	2.73	52.1	44.4	17.34
Sardinha et al. [34]	61.4	54.2	2.73	52.1	35.8	45.53
Sardinha et al. [34]	15.4	54.2	2.73	53.6	53.5	0.19
Sardinha et al. [34]	30.7	54.2	2.73	53.6	47.95	11.78
Sardinha et al. [34]	61.4	54.2	2.73	53.6	37.4	43.32
Singh et al. [20]	42.2	26.63	2.67	38.54	39.88	-3.36
Singh et al. [20]	63.3	26.63	2.67	38.54	41.35	-6.80
Singh et al. [20]	84.4	26.63	2.67	38.54	35.09	9.83
Singh et al. [20]	105.5	26.63	2.67	38.54	33.12	16.36
Singh et al. [20]	39.4	26.63	2.67	31.37	32.44	-3.30
Singh et al. [20]	59.1	26.63	2.67	31.37	33.65	-6.78
Singh et al. [20]	78.8	26.63	2.67	31.37	27.01	16.14
Singh et al. [20]	98.5	26.63	2.67	31.37	25.74	21.87
Singh et al. [20]	35.1	26.63	2.67	23.54	24.65	-4.50
Singh et al. [20]	52.65	26.63	2.67	23.54	23.08	1.99
Singh et al. [20]	70.2	26.63	2.67	23.54	20.42	15.28
Singh et al. [20]	87.75	26.63	2.67	23.54	20.03	17.52
Aliabdo [1]	20	83.22	2.5	39.93	37.3	7.05
Aliabdo [1]	30	83.22	2.5	39.93	38.4	3.98
Aliabdo [1]	40	83.22	2.5	39.93	38.8	2.91
Aliabdo [1]	60	83.22	2.5	39.93	34.6	15.40
Aliabdo [1]	20	83.22	2.5	48.73	51.76	-5.85
Aliabdo [1]	30	83.22	2.5	48.73	52.64	-7.43
Aliabdo [1]	40	83.22	2.5	48.73	53.12	-8.26

Author/s	Marble waste properties			Compressive strength		
	Content (kg)	CaO (%)	Specific Gravity	f_{cc} (MPa)	f_{cm} (MPa)	Δf_c (%)
Aliabdo [1]	60	83.22	2.5	48.73	48.44	0.60
Bostanci [35]	25	43.5	2.86	46	41.33	11.30
Bostanci [35]	50	43.5	2.86	46	38.44	19.67
Uysal & Yilmaz [25]	55	55.49	2.71	75.9	76.2	-0.39
Uysal & Yilmaz [25]	110	55.49	2.71	75.9	77.5	-2.06
Uysal & Yilmaz [25]	165	55.49	2.71	75.9	70.8	7.20
Kumar & Kumar [2]	22.28	55.09	2.63	33.18	34.67	-4.30
Kumar & Kumar [2]	44.56	55.09	2.63	33.18	35.85	-7.45
Kumar & Kumar [2]	66.84	55.09	2.63	33.18	30.22	9.79
Kumar & Kumar [2]	89.12	55.09	2.63	33.18	29.19	13.67

Table 3.
Collected data for the compressive strength estimation model.

Generally, a good fit is observed when the residuals of the estimation model are scattered randomly along the independent variable axis by representing positive and negative values. Furthermore, a potential outlier is identified when a value in standardized residuals exceeds approximately 3 [39, 40].

$$d_i = \frac{e_i}{\sqrt{MS_E}} \quad (4)$$

where MS_E is the mean squared error.

5.3 Compressive strength regression model

In this section, a prediction model for the compressive strength of concrete incorporating marble dust as a cement replacement will be discussed. In general, to overcome the dependency of the prediction model on the size of the testing specimen, the compressive strength of the control mixture will be used as an input to the model in which the output will represent the compressive strength of marble dust concrete specimen that has the same size as the inputted one. The analysis of variance of the predicted model is collected in **Table 5**, and the estimation model is depicted in (Eq. (6)) with an R^2 value of 0.9 representing a reasonably good fitting indicator. As recorded in **Table 5**, the p-value of the input parameters was below 5% for the cases of Marble dust content and its specific gravity and the control compressive strength. In comparison, the CaO content had a higher value. Although this observation does not give a very solid conclusion to what is the most influencing parameter on the compressive strength, it can supply a rough idea that marble specific gravity has a higher effect on the compressive strength than the marble content and its CaO. A similar conclusion is derived using the simple linear regression approach, **Figure 6**, between these factors and the change in the compressive strength (Eq. (5)), where higher R^2 values refer to the better correlation and consequently the more considerable effect.

$$\Delta f_c = \frac{f_{cc} - f_{cm}}{f_{cc}} \times 100 \quad (5)$$

$$f_{cm} = 51.6 + 1.0123 f_{cc} - 0.0468 MD_c - 0.0473 MD_{CaO} - 18.6 MD_{SG} \quad (6)$$

Author/s	Marble waste properties			Flexural strength		
	Content (kg)	CaO (%)	Specific Gravity	f_{tc} (MPa)	f_{tm} (MPa)	Δf_t (%)
Ergün [3]	15	51.7	2.68	5.3	5.3	0.00
Ergün [3]	22.5	51.7	2.68	5.3	5.1	3.92
Ergün [3]	30	51.7	2.68	5.3	5	6.00
Rana et al. [32]	20.25	65.2	2.87	6.55	6.3	3.97
Rana et al. [32]	40.5	65.2	2.87	6.55	6.25	4.80
Rana et al. [32]	60.75	65.2	2.87	6.55	5.44	20.40
Rana et al. [32]	81	65.2	2.87	6.55	5.14	27.43
Rana et al. [32]	101.25	65.2	2.87	6.55	5	31.00
Singh et al. [20]	42.2	26.63	2.67	7.8	8.17	-4.53
Singh et al. [20]	63.3	26.63	2.67	7.8	8.15	-4.29
Singh et al. [20]	84.4	26.63	2.67	7.8	7.32	6.56
Singh et al. [20]	105.5	26.63	2.67	7.8	7.1	9.86
Singh et al. [20]	39.4	26.63	2.67	6.83	7	-2.43
Singh et al. [20]	59.1	26.63	2.67	6.83	7.02	-2.71
Singh et al. [20]	78.8	26.63	2.67	6.83	6.16	10.88
Singh et al. [20]	98.5	26.63	2.67	6.83	6.08	12.34
Singh et al. [20]	35.1	26.63	2.67	5.7	5.803	-1.77
Singh et al. [20]	52.65	26.63	2.67	5.7	5.905	-3.47
Singh et al. [20]	70.2	26.63	2.67	5.7	5.408	5.40
Singh et al. [20]	87.75	26.63	2.67	5.7	5.12	11.33
Kumar & Kumar [2]	22.28	55.09	2.63	5.33	5.43	-1.84
Kumar & Kumar [2]	44.56	55.09	2.63	5.33	5.63	-5.33
Kumar & Kumar [2]	66.84	55.09	2.63	5.33	5.73	-6.98
Kumar & Kumar [2]	89.12	55.09	2.63	5.33	4.7	13.40

Table 4.
Collected data for the flexural strength estimation model.

Analysis of variance					
Source	DF	Adj SS	Adj MS	F-Value	P-Value
Regression	4	6838.16	1709.54	98.05	0
Marble dust content	1	86.85	86.85	4.98	0.031
CaO	1	22.52	22.52	1.29	0.262
Specific gravity of marble dust	1	149.57	149.57	8.58	0.005
Control compressive strength	1	5249.81	5249.81	301.11	0
Error	44	767.15	17.44		
Total	48	7605.31			

Table 5.
Analysis of variance for the compressive strength model.

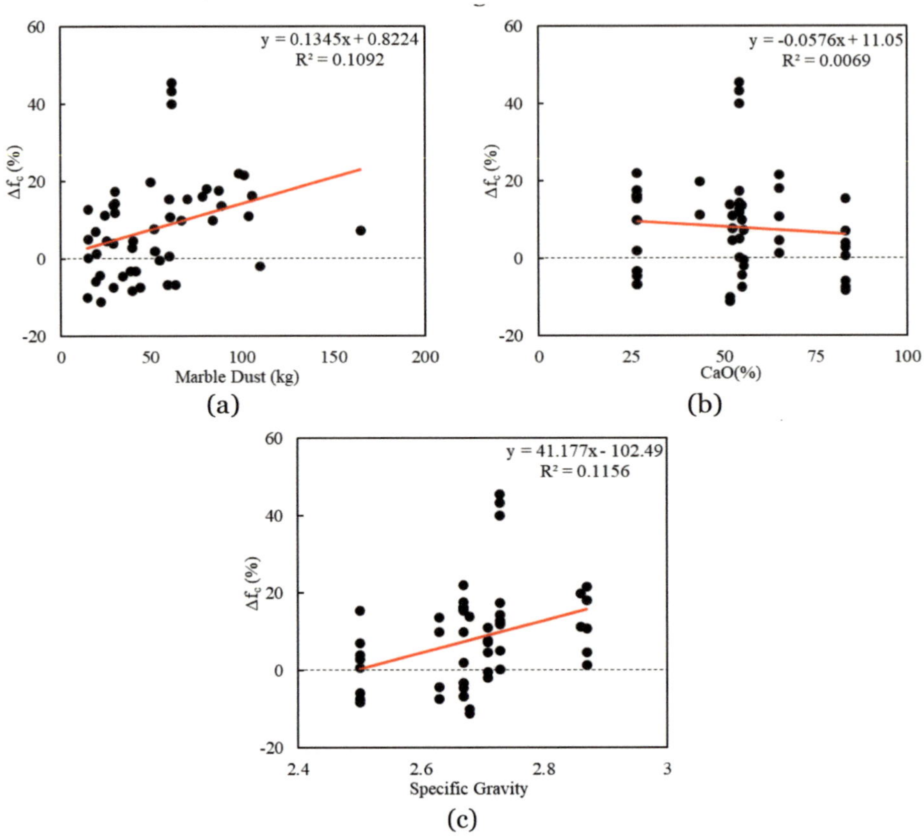

Figure 6.
Influence of marble dust properties on the change in the compressive strength (a) marble dust conten, (b) CaO in the marble dust, (c) specific gravity of the marble dust.

The performance of the interpolation model is represented in **Figure 7**. It is seen that the regression line between the measured and predicted values is mainly lying over the equality line, and the points are distributed all over it, which represents a good fitting model. The residuals of this predictor are seen in **Figure 8**, and these

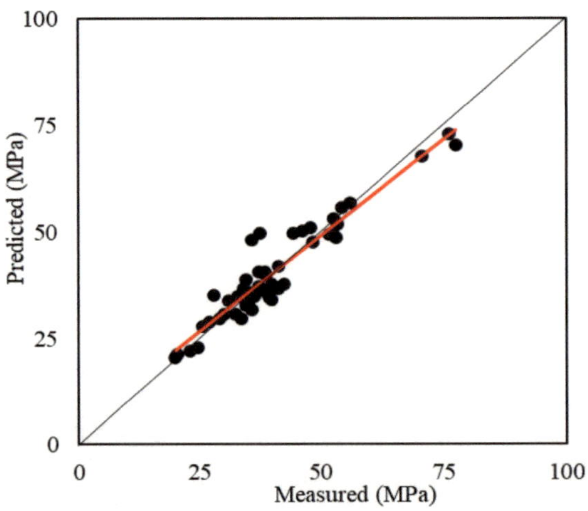

Figure 7.
Performance of the proposed compressive strength prediction model.

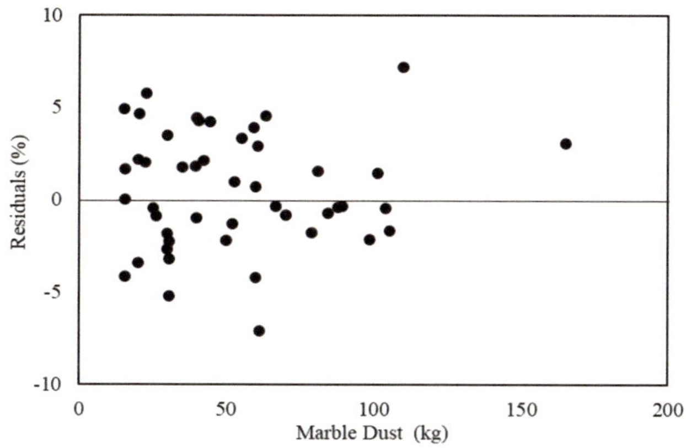

Figure 8.
Residuals obtained from the compressive strength prediction model.

values follow the basic requirements of being randomly scattered along the dependent variable axis, which is the marble dust content, in this case, meaning that the proposed model is appropriate for the given dataset.

The standardized residuals were obtained to determine the outliers, **Figure 9**, which their values more than three. Hence, it can be noticed that almost no outliers has occurred in this study, which indicates a good fitting capability. Another check on the prediction model is shown in **Figure 10**, in which the distribution on the predicted values is compared to this of the experimental one. Indeed, the histograms are slightly changed in the case of estimation; however, the general distribution is conserved after prediction.

Finally, it can be observed that the proposed prediction model delivers a useful capability for estimating the compressive strength of concrete mixtures utilizing marble dust as a replacement of cement.

5.4 Flexural strength regression model

The analysis of variance of the predicted model is summarized in **Table 6**, and the prediction model is stated in Eq. (8) with an R^2 value of 0.93 representing a

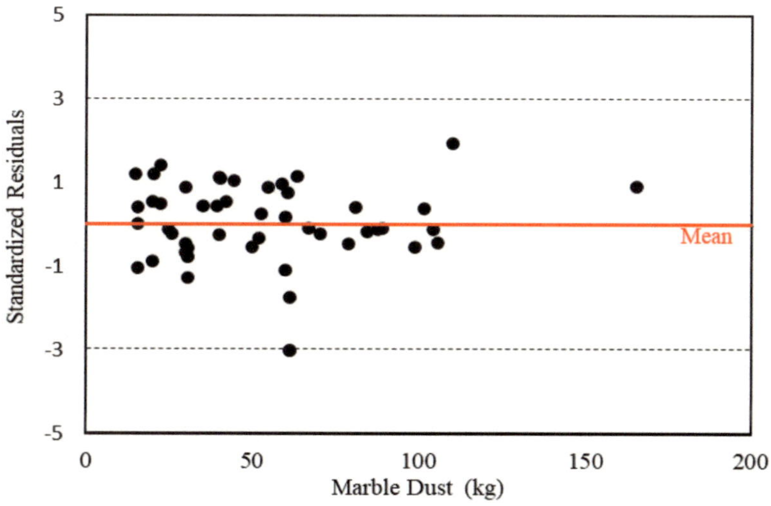

Figure 9.
Standardized residuals of the proposed compressive strength prediction model.

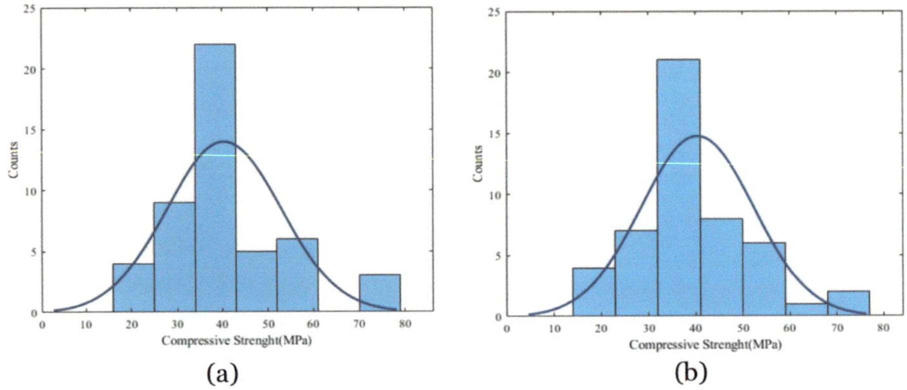

Figure 10.
Histogram and normal distribution for the compressive strength (a) experimental dataset, and (b) predicted values.

Analysis of variance					
Source	**DF**	**Adj SS**	**Adj MS**	**F-Value**	**P-Value**
Regression	4	20.66	5.16	61.27	0
Marble dust content	1	2.77	2.77	32.86	0
CaO	1	0.00	0.00	0.04	0.837
Specific gravity of marble dust	1	1.00	1.00	11.85	0.003
Control compressive strength	1	11.62	11.62	137.85	0
Error	19	1.60	0.08		
Total	23	22.26			

Table 6.
Analysis of variance for the flexural strength model.

reasonable good fitting indicator. In similar to the compressive strength, the specific gravity of the marble dust has the most influence on the flexural strength as compared to the CaO and effect. Also, the same conclusion can be acquired using the simple linear regression approach, **Figure 11**, between these factors and the change in the flexural strength (Eq. (7)).

$$\Delta f_t = \frac{f_{tc} - f_{tm}}{f_{tc}} \times 100 \tag{7}$$

$$f_{tm} = 10.29 + 1.1452 f_{tc} - 0.0137 MD_c - 0.00133 MD_{CaO} - 3.93 MD_{SG} \tag{8}$$

The performance of the estimation model is described in **Figure 12**. It can be seen that good fitting is obtained for the given dataset with a reasonably high R^2 value in comparison to the compressive strength. The residuals of this model are placed in **Figure 13**. It can be discovered that these values are randomly scattered along the marble dust content axis, which represents that the proposed model fits the given dataset.

The standardized residuals approach was used to investigate the occurrence of potential outliers, as indicated in **Figure 14**. In general, it can be observed that no points have exceeded 3, which means no outliers have existed. In addition, the

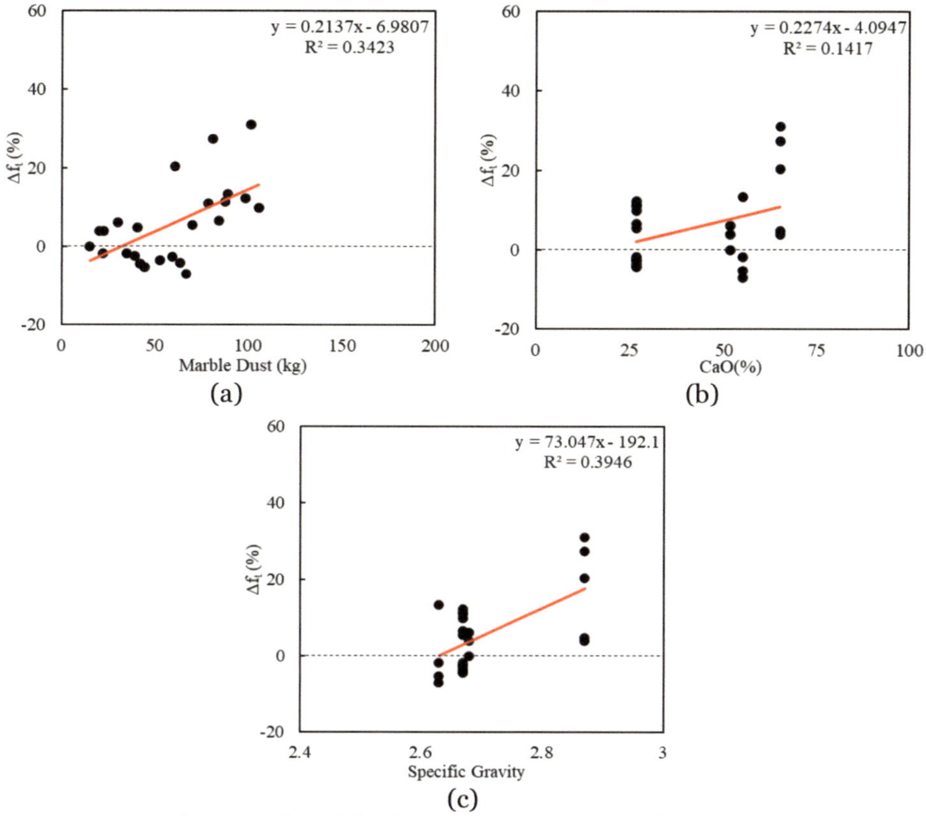

Figure 11.
Influence of marble dust properties on the change in the flexural strength (a) marble dust conten, (b) CaO in the marble dust, (c) specific gravity of the marble dust.

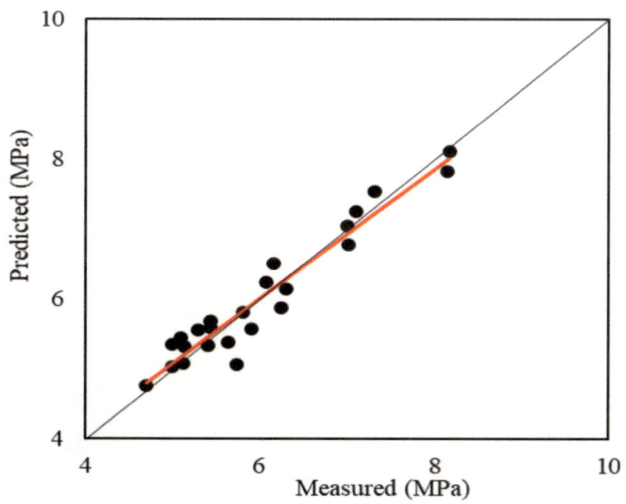

Figure 12.
Performance of the proposed flexural strength prediction model.

distribution of the predicted values as compared to the experimental one can be found in **Figure 15**. A slight variation can be seen, although the behavior is generally conserved in both cases, reflecting a suitable fitting capability for the dataset.

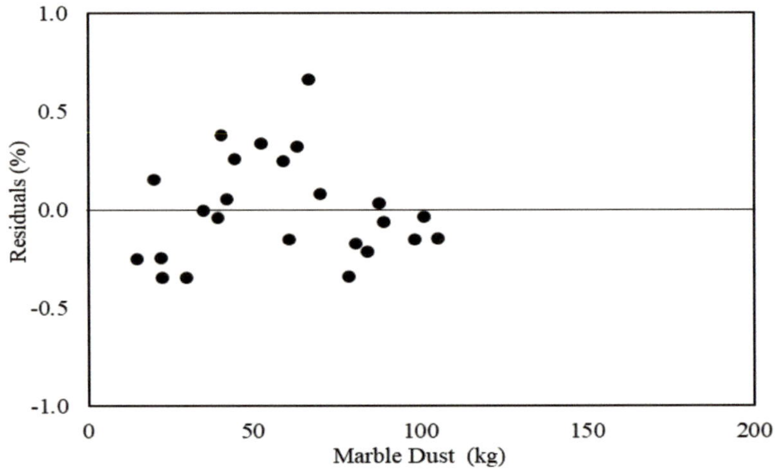

Figure 13.
Residuals obtained from the flexural strength prediction model.

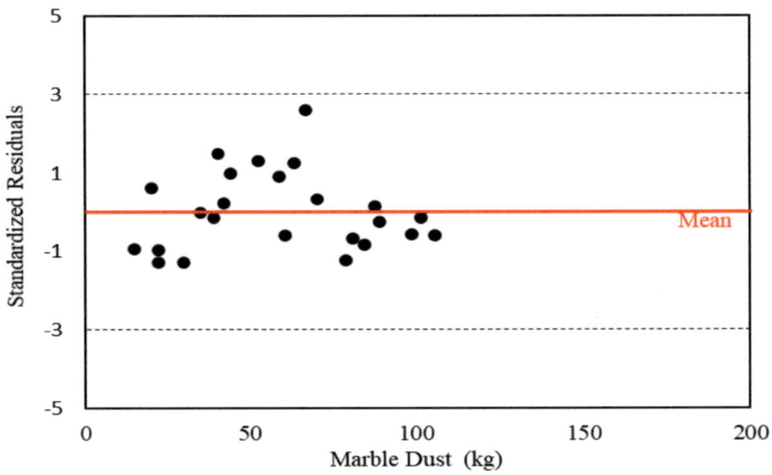

Figure 14.
Standardized residuals of the proposed flexural strength prediction model.

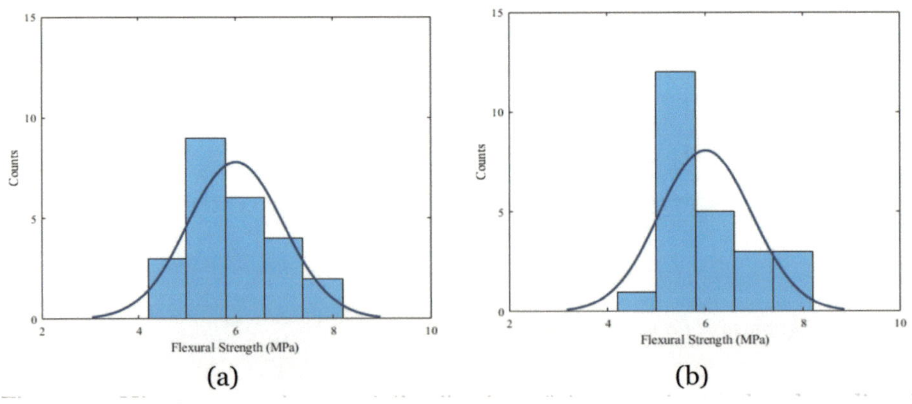

Figure 15.
Histogram and normal distribution for the flexural strength (a) experimental dataset, and (b) predicted values.

6. Future trends

Many efforts were focused on the influence of marble dust on the properties of several types of cement-based materials. However, it was observed that the literature is still in need of some comprehensives studies that can help the scientific community to understand the influence of using marble dust as a filler in sustainable concrete mixtures that incorporate recycled aggregates such as plastic or rubber. Such a mixture might provide a promising sustainable solution for ultimate waste management plans in developing countries where such materials are highly available. Another research gap is mainly related to the dynamic properties of mixtures incorporating marble dust, as such studies are minimal. Moreover, it is essential to display some numerical studies that can propose prediction models based on the marble dust content in the concrete mixture.

7. Conclusion

This chapter has focused on briefly reviewing the utilization and mechanical properties of cement-based materials incorporating marble dust with emphasis on concrete mixtures with marble wastes cement replacement. In addition, it aimed to propose two estimation models using multiple regression analysis for the compressive and flexural strengths of marble dust concrete. On the base of the statements above, the following points are drawn:

- It is quite challenging to narrow down the ranges of marble dust properties and to standardize them due to the massive variate in the origin of the rocks being processed while obtaining this material.

- Marble dust can be used as a filler material in concrete to improve the microstructure of the mix.

- Up to a certain replacement ratio, generally considered as 10% in several studies, the incorporation of marble waste can positively influence the compressive strength capacity of the mixture.

- Several potential applications of marble dust in the construction industry have already been considered in the literature, including its utilization as a filler in cement-based materials, a partial replacement of concrete constituents, and a substitute of limestone in various industrial applications.

- The specific gravity of marble dust can be considered as one of the main characteristics that affect the strength properties of the investigated concrete mixtures.

- The proposed estimation models can reliably be used to predict the compressive and flexural strengths of concrete utilizing marble dust as a partial replacement of cement.

Further research efforts are still needed in this field to cover some of the gaps in the literature on the behavior of recycled aggregate concrete incorporating this material to develop the understanding of both scientists and engineers working in the construction industry. It is also recommended to comprehensively study the influence of marble dust chemical properties on the performance of the produced cement-based material can to come up with detailed mathematical relationships similar to the ones presented in this study due to the inconsistency in the experimental results due to the wide variety in the natural waste properties.

Author details

Ahed Habib[1]* and Maan Habib[2]

1 Eastern Mediterranean University, Famagusta, North Cyprus, via Mersin 10, Turkey

2 Al-Balqa Applied University, Amman, Jordan

*Address all correspondence to: ahed.habib@cc.emu.edu.tr

References

[1] A. A. Aliabdo, M. Abd Elmoaty and E. M. Auda, "Re-use of waste marble dust in the production of cement and concrete," Construction and building materials, vol. 50, pp. 28-41, 2014

[2] R. Kumar and S. K. Kumar, "Partial replacement of cement with marble dust powder," International Journal of Engineering Research and Applications, vol. 5, no. 8, pp. 106-114, 2015

[3] A. Ergün, "Effects of the usage of diatomite and waste marble powder as partial replacement of cement on the mechanical properties of concrete," Construction and building materials, vol. 25, no. 2, pp. 806-812, 2011

[4] R. Rodrigues, J. De Brito and M. Sardinha, "Mechanical properties of structural concrete containing very fine aggregates from marble cutting sludge," Construction and Building Materials, vol. 77, pp. 349-356, 2015

[5] O. Gencel, C. Ozel, F. Koksal, E. Erdogmus, G. Martínez-Barrera and W. Brostow, "Properties of concrete paving blocks made with waste marble," Journal of cleaner production, vol. 21, no. 1, pp. 62-70, 2012

[6] V. Corinaldesi, G. Moriconi and T. R. Naik, "Characterization of marble powder for its use in mortar and concrete," Construction and building materials, vol. 24, no. 1, pp. 113-117, 2010

[7] E. T. Tunc, "Recycling of marble waste: A review based on strength of concrete containing marble waste," Journal of environmental management, vol. 231, pp. 86-97, 2019

[8] M. Galetakis and A. Soultana, "A review on the utilisation of quarry and ornamental stone industry fine by-products in the construction sector," Construction and Building Materials, vol. 102, pp. 769-781, 2016

[9] H. Y. Aruntaş, M. Gürü, M. Dayı and I. Tekin, "Utilization of waste marble dust as an additive in cement production," Materials & Design, vol. 31, no. 8, pp. 4039-4042, 2010

[10] B. Demirel, "The effect of the using waste marble dust as fine sand on the mechanical properties of the concrete," International journal of physical sciences, vol. 5, no. 9, pp. 1372-1380, 2010

[11] H. Ş. Arel, "Recyclability of waste marble in concrete production," Journal of Cleaner Production, vol. 131, pp. 179-188, 2016

[12] H. E. Elyamany, M. Abd Elmoaty and B. Mohamed, "Effect of filler types on physical, mechanical and microstructure of self compacting concrete and Flow-able concrete," Alexandria Engineering Journal, vol. 53, no. 2, pp. 295-307, 2014

[13] M. G. H. Education, Dictionary of Geology & Mineralogy, McGraw-Hill , 2003

[14] P. Julphunthong and P. Joyklad, "Utilization of Several Industrial Wastes as Raw Material for Calcium Sulfoaluminate Cement," Materials, vol. 12, no. 20, p. 3319, 2019

[15] B. Demirel and K. E. Alyamaç, "Waste marble powder/dust," in Waste and Supplementary Cementitious Materials in Concrete, Woodhead Publishing, 2018, pp. 181-197

[16] M. Gesoğlu, E. Güneyisi, M. E. Kocabağ, V. Bayram and K. Mermerdaş, "Fresh and hardened characteristics of self compacting concretes made with combined use of marble powder, limestone filler, and fly ash," Construction and Building Materials, vol. 37, pp. 160-170, 2012

[17] K. Vardhan, S. Goyal, R. Siddique and M. Singh, "Mechanical properties and microstructural analysis of cement mortar incorporating marble powder as partial replacement of cement," Construction and Building Materials, vol. 96, pp. 615-621, 2015

[18] D. K. Ashish, "Concrete made with waste marble powder and supplementary cementitious material for sustainable development," Journal of cleaner production, vol. 211, pp. 716-729, 2019

[19] N. Bilgin, H. A. Yeprem, S. Ö. N. M. E. Z. Arslan, A. Bilgin, E. Günay and M. Marşoglu, "Use of waste marble powder in brick industry," Construction and Building Materials, vol. 29, pp. 449-457, 2012

[20] M. A. Singh and D. Bhunia, "An investigation on effect of partial replacement of cement by waste marble slurry," Construction and Building Materials, vol. 134, pp. 471-488, 2017

[21] L. G. Li, Z. H. Huang, Y. P. Tan, A. K. H. Kwan and H. Y. Chen, "Recycling of marble dust as paste replacement for improving strength, microstructure and eco-friendliness of mortar," Journal of Cleaner Production, vol. 210, pp. 55-65, 2019

[22] T. Uygunoğlu, İ. B. Topçu and A. G. Çelik, "Use of waste marble and recycled aggregates in self-compacting concrete for environmental sustainability," Journal of cleaner production, vol. 84, pp. 691-700, 2014

[23] M. Uysal and M. Sumer, "Performance of self-compacting concrete containing different mineral admixtures," Construction and Building materials, vol. 25, no. 11, pp. 4112-4120, 2011

[24] I. B. Topcu, T. Bilir and T. Uygunoğlu, "Effect of waste marble dust content as filler on properties of self-compacting concrete," Construction and building Materials, vol. 23, no. 5, pp. 1947-1953, 2009

[25] M. Uysal and K. Yilmaz, "Effect of mineral admixtures on properties of self-compacting concrete," Cement and Concrete Composites, vol. 33, no. 7, pp. 771-776, 2011

[26] A. Talah, F. Kharchi and R. Chaid, "Influence of marble powder on high performance concrete behavior," Procedia Engineering, vol. 114, pp. 685-690, 2015

[27] F. Gameiro, J. De Brito and D. C. da Silva, "Durability performance of structural concrete containing fine aggregates from waste generated by marble quarrying industry," Engineering Structures, vol. 59, pp. 654-662, 2014

[28] N. Toubal Seghir, M. Mellas, Ł. Sadowski, A. Krolicka, A. Żak and K. Ostrowski, "The utilization of waste marble dust as a cement replacement in air-cured mortar," Sustainability, vol. 11, no. 8, p. 2215, 2019

[29] A. O. Mashaly, B. A. El-Kaliouby, B. N. Shalaby, A. M. El–Gohary and M. A. Rashwan, "Effects of marble sludge incorporation on the properties of cement composites and concrete paving blocks," Journal of Cleaner Production, vol. 112, pp. 731-741, 2016

[30] M. Singh, A. Srivastava and D. Bhunia, "Potential applications of marble dust in industrial use by characterization techniques—A review," International Journal of Advanced Structures and Geotechnical Engineering, vol. 5, no. 3, pp. 99-106, 2016

[31] M. J. Munir, S. M. S. Kazmi and Y. F. Wu, "Efficiency of waste marble powder in controlling alkali–silica reaction of concrete: A sustainable

approach," Construction and Building Materials, vol. 154, pp. 590-599, 2017

[32] A. Rana, P. Kalla and L. J. Csetenyi, "Sustainable use of marble slurry in concrete," Journal of Cleaner Production, vol. 94, pp. 304-311, 2015

[33] N. M. Soliman, "Effect of using marble powder in concrete mixes on the behavior and strength of RC slabs," International Journal of Current Engineering and Technology, vol. 3, no. 5, pp. 1863-1870, 2013

[34] M. Sardinha, J. de Brito and R. Rodrigues, "Durability properties of structural concrete containing very fine aggregates of marble sludge," Construction and building materials, vol. 119, pp. 45-52, 2016

[35] S. C. Bostanci, "Use of waste marble dust and recycled glass for sustainable concrete production," Journal of Cleaner Production, vol. 251, 2020

[36] S. Fedorov, "GetData graph digitizer (2.26)," www.getdata-graph-digitizer.c om, 2013

[37] D. J. Olive, Multiple Linear and 1D Regression, 2010

[38] C. H. Achen, Interpreting and using regression, vol. 29, Sage, 1982

[39] D. C. Montgomery, E. A. Peck and G. G. Vining, Introduction to linear regression analysis, John Wiley & Sons, 2012

[40] D. C. Montgomery, Design and analysis of experiments, John Wiley & sons, 2012

Chapter 11

Applications of Cement in Pavement Engineering

Sarella Chakravarthi, Galipelli Raj Kumar and Sabavath Shankar

Abstract

Recycled materials primarily Reclaimed Asphalt Pavement (RAP), and Recycled Concrete Aggregate (RCA) are produced from pavement rehabilitation and construction-demolition activities. Generally, these materials are utilized for landfills, parking lots, shoulders, and other places that are not environmentally friendly. The top layers of the pavement and concrete structures are constructed using superior qualities of aggregates that satisfy the specification. During their service life, the aggregates present in these structures undergo deterioration due to environmental and traffic factors. After reaching the end of their service life, the deteriorated structures are dismantled and considered as waste. Nevertheless, these recycled materials will have some retain value which can be used in different layers of the pavements in different percentages. The reuse of these materials in place of conventional aggregates preserves the environment and become a sustainable construction practice. Further, the direct utilization of these materials in the pavements may not satisfy the mechanical characteristics. To fulfill these gaps, cement stabilization of recycled materials is the best option. With this background, the proposed book chapter will highlight the usage of cement in pavement application, and a few types of research works carried in cement treated pavement layers will be discussed in a detailed and scientific manner.

Keywords: cement concrete pavement, granular layers, cement treated bases and performance of pavements

1. Introduction

Recycled materials primarily Reclaimed Asphalt Pavement (RAP), and Recycled Concrete Aggregate (RCA) are produced from pavement rehabilitation and construction-demolition activities. Generally, these materials are utilized for landfills, parking lots, shoulders, and other places that are not environmentally friendly. The top layers of the pavement and concrete structures are constructed using superior qualities of aggregates that satisfy the specification. During their service life, the aggregates present in these structures undergo deterioration due to environmental and traffic factors. After reaching the end of their service life, the deteriorated structures are dismantled and considered as waste. However, these recycled materials have some retain value which can be used in different layers of the pavements in different percentages. The reuse of these materials in place of conventional aggregates preserves the environment and become a sustainable construction practice. However, the direct utilization of these materials in the pavements may not achieve acceptable mechanical characteristics. To fulfill these mechanical properties with recycled materials, stabilization is the best option left to the engineers.

Several stabilization techniques are involved around the world to provide adequate strength to the weak bases or soil materials. The stabilizers include lime, asphalt emulsion, fly ash, and cement are widely used to improve the mechanical properties of recycled materials as a base or subbase courses in pavements. Cement stabilization is advantageous because of rapid gain in strength and its easy availability in the market. To understand the mechanical properties of the recycled materials with cement stabilization, a laboratory study is carried using RAP and RCA in different proportions with Natural Aggregates (NA). The compaction characteristics, Unconfined Compressive Strength (UCS), Indirect Tensile Strength (ITS), and Modulus of Elasticity (E) tests were conducted to assess the performance.

Apart from the recycled aggregates, several hazardous and industrial wastes from the production plants like Electrolyte Manganese Residue (EMR), Red mud, slag, and glass can be efficiently stabilized using cement and can be used in the pavements as a base. Zang et al. (2019) proved that the stabilization of the EMR and Red mud in the road bases achieved adequate strength and make it environmentally friendly [1]. The replacement of conventional aggregates with 50% steel slag stabilized with 4% cement content achieved maximum strength and stiffness along with other economic benefits [2]. At the same time, the use of cement treated recycled glass up to 30% along with the other recycled materials achieved required strength properties [3]. Besides, it is estimated from the study that 26-32% of cost savings with the cement stabilization of the recycled aggregates [4].

The usage of cement in pavement construction is considerable. There are several applications in the construction field that includes bridges, tunnels, safety barriers, pavements, and sound barriers. The benefits of cement-treated bases include high bearing capacity and increased stiffness, and lower deformation under loads. The following are the potential advantages of cement in various construction fields which are listed below.

- Airports for parking aprons, taxiway and runway take-off
- Parking areas for heavy vehicles
- Heavy-duty industrial floors
- Usage of cement and concrete in bridge decks
- Subgrade soil stabilization with cement
- Soil stabilization with lime and cement
- Recycling of pavement with treated and bases and subbases
- Construction of dry lean cement concrete layer
- Debonding layer over stabilized cemented bases
- Construction of concrete pavement
- Construction of interlocking bloc pavement s
- Cell filled pavements.

With this background, the application of the cement in the field of pavement engineering as a base and sub-base layer is presented. This includes evaluation in terms of mechanical properties at various cement stabilization levels and finally compared with the available specifications.

2. Literature review

The potential use of recycled aggregates in pavement bases is investigated around the world. Their utilization in the pavement is limited due to inferior physical and mechanical properties. Reclaimed asphalt pavement (RAP) materials used in the pavements has concerns with the reduction in the strength, more significant permanent deformation, poor distribution of stresses, and durability issues and recommended for stabilization [5, 6]. Besides, the use of recycled concrete aggregates (RCA) above the water table is recommended [7] due to the concerns of groundwater contamination and durability issues [8, 9]. Studies suggested that the properties of the recycled materials can be enhanced with the addition of additives or blending with superior quality materials [10]. The stiffness of the base increases with the cement stabilization, which reduces the deflections and increases the pavement life at higher traffic loads and serves better than NA bases [8, 9]. With this background, there is a stressing need for chemical stabilization of recycled materials to improve their mechanical properties. Some of the previous studies were presented in **Table 1**, which shows the benefits of the cement stabilization on various recycling materials.

Author, year	Conclusions
Arulrajah et al. 2020 [11]	Stabilization of construction and demolition wastes containing a little amount of polyethylene terephthalate with 3% cement satisfied the requirements of pavement base and subbases.
Kasu et al. 2020 [4]	Cement content in the mix has a significant influence on the mechanical and durability properties compared with recycled aggregate contents.
Arshad, 2020 [12]	Addition of cement content in the range of 1.5-4.5% improved the mechanical properties and reduced the strains in the recycled aggregate blends
Yan et al. 2020 [13]	The addition of recycled aggregate content from the construction and demolition waste into the stabilized bases improves the mechanical and durability properties. However, the residual strength is maximum at 30% of the recycled aggregate content in the mix.
Chakravarthi et al. 2019 [14]	Concluded that cement stabilization of the RCA is more pronounced than RAP and improvement in the mechanical properties is observed.
Faysal et al. 2016 [15]	Strength and stiffness characteristics are significantly improved with cement content.
LaHucik et al. 2016 [16]	Cement stabilization on a small scale for the bases with recycled materials like RAP and Quarry by-products is feasible for no freeze zones.
Behiry, 2013 [17]	Revealed that the performance of cement treated recycled materials depends on cement content, curing time and dry density.
Xuan et al. 2012 [18]	Stated that an increase in the cement content and decreasing the masonry content improves the strength and modulus of the bases.
Taha et al. 2002 [19]	Replacement of cement stabilization of RAP-virgin aggregate mixes in place of conventional bases considered a viable alternative.

Table 1.
Previous studies on cement stabilization of recycled bases.

3. Objectives and methodology of the study

The current study aims at laboratory evaluation of the cement stabilized bases consists of recycled materials like RAP and RCA as a partial and full replacement with natural Aggregates (VA) with the following goals.

- To determine the influence of the cement content and recycled aggregates content on the stabilized base mixes in terms of various mechanical properties.
- To optimize the cement content, recycled aggregates, and VA combination for road bases based according to the standard specifications.
- Stiffness characterization of the given recycled aggregates and VA combinations at different stabilization levels.

The required materials such as conventional aggregate, RAP, RCA, and ordinary Portland cement of 53 grade are collected locally. Aggregate gradations for the test mixtures were determined and compared with the requirements following the Ministry of Road Transport and Highways (MoRTH, 2013) specifications, Government of India (GoI) [20]. Initially, bitumen was extracted from the collected RAP material, followed by physical tests like Sieve analysis, impact test, Flakiness, and Elongation index were performed on RAP and RCA according to the standards as shown in **Figure 1**. From the physical properties, the flakiness and elongation index for RAP is higher than the specification limits, and the water absorption of the RCA is more than 2%. The flakiness and elongation index of RAP aggregates is due to the formation of fracture surfaces during its service life. The crushing process and the water absorption of the RCA are more compared with the VA because of the presence of cement mortar around its surface. Although the recycled materials did not satisfy the required specifications, they are used in the study. This is the main motivation of the study to improve their mechanical properties through the process of stabilization. After physical characterization, the materials are blended in the ratio of RAP/RCA and conventional aggregate content (0/100, 25/75, 50/50, 75/25) as shown in **Figure 2** with an increment of cement from 0%, 2%, 4%, 6%. The next step involves the determination of compaction characteristics of the mixes such as Optimum moisture content

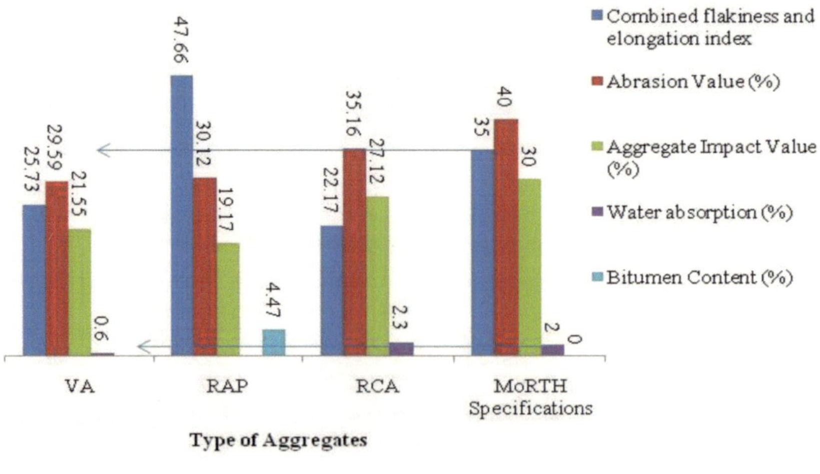

Figure 1.
Physical properties of aggregates.

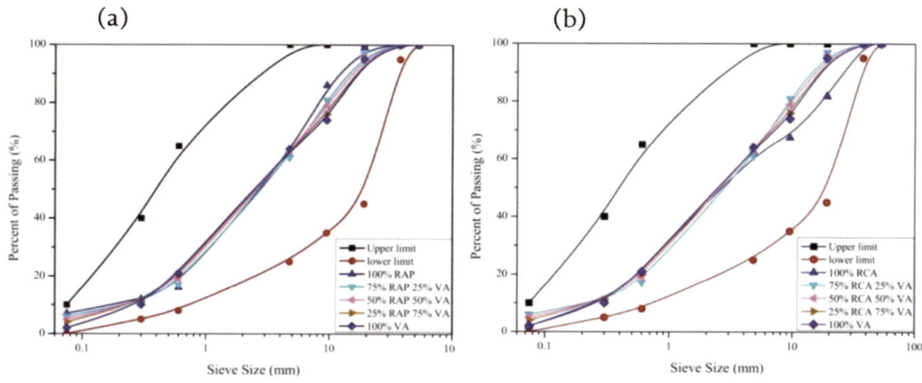

Figure 2.
(a) Gradation curve for RAP blends; (b) gradation curve for RCA blends.

(OMC) and maximum dry density (MDD) using the Modified Proctor Test. The obtained OMC values from the modified Proctor test are used in the preparation of the samples for further tests and cured for 7 days as suggested by the researchers. The strength parameters like UCS, ITS, and stiffness parameters like Modulus of elasticity and Resilient Modulus were evaluated for the specimens prepared at OMC.

4. Compaction characteristics

Modified Proctor's test was performed according to the Indian Standards (IS 2720-part 8-1985) on all the mixes proportions of RAP and RCA with VA. The test is repeated three times to check the repeatability and accuracy. Modified Proctor compaction is achieved using a hammer of weight 4.5 kg falling from a height of 457 mm in the mold of dimensions 102 mm in diameter and 127 mm in height. All particle sizes greater than 19 mm were replaced with the same amount of particles less than 19 mm from the mix as a mold correction.

The obtained OMC and MDD results for different blending mixes are shown in **Tables 2** and **3**. It is revealed that the required amount of OMC is reduced with the increase in the percentage of RAP due to the low moisture absorption capacity of bitumen coated RAP. While the MDD is low at 100% RAP and increases with VA content due to the low specific gravity of the RAP aggregates which agrees with the previous research study [21] and MDD increases with the cement content. The addition of the cement to the mix improves the compaction capacity due to good aggregate packing. In the case of RCA blends, there is no proper trend observed with the addition of RCA due to the indifferences in mortar on the surface of RCA. The same is verified by conducting a water absorption test on two samples of the

% of cement	100% RAP		75% RAP		50% RAP		25% RAP	
	OMC (%)	MDD (g/cc)	OMC (%)	MDD (g/cc)	OMC (%)	MDD (g/cc)	OMC (%)	MDD (g/cc)
0	7.06	1.93	7.44	2.03	7.61	2.13	7.16	2.20
2	7.13	2.08	7.52	2.08	7.72	2.14	7.38	2.22
4	7.24	2.11	7.64	2.10	7.89	2.15	7.67	2.22
6	7.45	2.21	7.88	2.11	7.95	2.16	8.09	2.23

Table 2.
Optimum moisture content and maximum dry density results for RAP blends.

% of cement	100% RCA		75% RCA		50% RCA		25% RCA	
	OMC (%)	MDD (g/cc)	OMC (%)	MDD (g/cc)	OMC (%)	MDD (g/cc)	OMC (%)	MDD (g/cc)
0	9.7	2.14	9.2	2.13	12.3	2.02	9.44	2.13
2	8.2	2.1	10.2	1.98	7.93	2.16	10.06	2.2
4	11.1	2.06	9.2	2.01	9.28	2.17	9.44	2.22
6	9.5	2.09	11.5	1.97	10.92	2.2	10.35	2.25

Table 3.
Optimum moisture content and maximum dry density results for RCA blends.

same gradation with the difference in the mortar presence. This variation of mortar percentage in the sample causes significant variations in the OMC and MDD at high percentages of RCA in the mix (100%RCA and 75% RCA). The MDD values of RAP and RCA blends ranges between 1.93 and 2.25 g/cc.

5. Unconfined compressive strength test (UCS) and elastic modulus

The Unconfined Compressive Strength (UCS) is used to determine the bonding strength or cohesion of a stabilized material. The samples are prepared at corresponding OMC according to ASTM D 1632. The dimensions of the cylindrical mold are of size 101.6 mm diameter, and 200 mm height is chosen. The samples are compacted and cured in closed plastic bags to prevent the escape of moisture. They tested at the end of the 7 days curing period. The UCS is calculated from the maximum load at the failure divided by the cross-sectional area of each specimen gives the compressive strength (**Figure 3**).

There is a surge in the UCS with cement content, as observed in **Figure 4**. Besides, the rate of increase in strength declines with the addition of recycled aggregates content. Higher RAP content slows down the rate of strength gain in the mixtures. For example, the mixes with increased recycled aggregates content have low strength at given cement content. The increase in asphalt coated surface area requires more amount of the stabilizing agent to form bonds with the other aggregates. Besides, the RCA stabilized bases do not show any particular trend with RCA content.

Figure 3.
Sample testing of UCS.

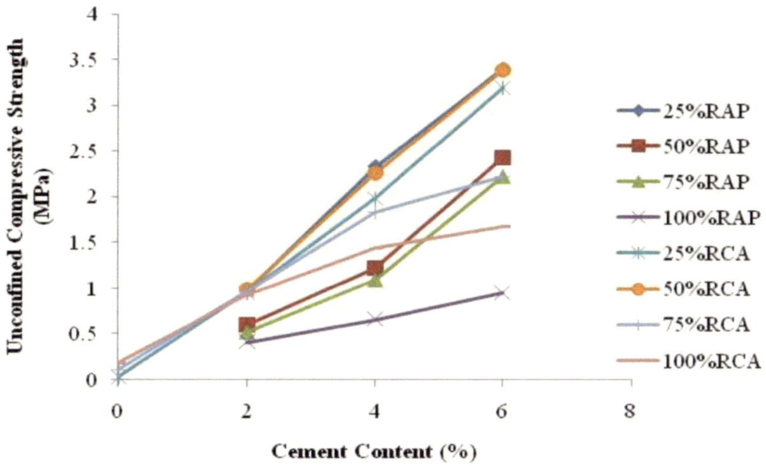

Figure 4.
Unconfined compressive strengths of RAP - VA and RCA - VA blends of varying cement percentages.

Further, 50% of RCA shows higher strength irrespective of the cement content; this is due to the better interlocking of RCA with the NA, which increases the strength. At 6% cement content, all the mixes exhibit higher strengths. The obtained results are compared with the low volume road standards for cement-treated bases, as shown in **Table 4**.

From the observations, the majority of the blends at 4% cement satisfied the specifications as a subbase layer for low volume roads. However, RAP/RCA blends with 25% RAP and 6% cement content have UCS of 3.4 MPa/3.19 MPa and 50% RCA with 6% cement with UCS of 3.37 MPa satisfied the Ministry of Rural Development (MoRD) specification, i.e., 2.76 MPa and can be used as a base layer for low volume roads and as a subbase layer for high volume roads. To extend their utilization as a base layer in the high-volume roads requires an increase in stabilization levels and curing period.

Further, RCA blends show more strength compared with RAP blends. For instance, at constant cement content, 100% RCA mix show almost double strength compared with 100% RAP mix. The reason behind this phenomenon is due to the existence of a strong bond between RCA, VA, and cement. Besides, the mortar which is coated with the RCA aggregates contributes to the development of strength. The RCA blends without stabilization exhibited retained strength at 7 days of curing. This clearly explains the self-cementing property of the RCA in agreement with the previous studies where the mortar present in the RCA helps in bonding [23].

Besides, the untreated RAP mixes are weak and collapsed while removing from the split mold, which represents the weak bonding between the aggregates. The strength development in the blends depends on the blended aggregate proportions, stabilization level, and the residual cement present in the existing RCA. However,

UCS (MPa)			
Low volume road (traffic <2 msa) [22]		**High volume road (traffic >2 msa) [20]**	
Sub-base	**Base**	**Sub-base**	**Base**
1.70	2.76	1.5-3.0	4.5-7.0

Table 4.
UCS as per Indian specifications.

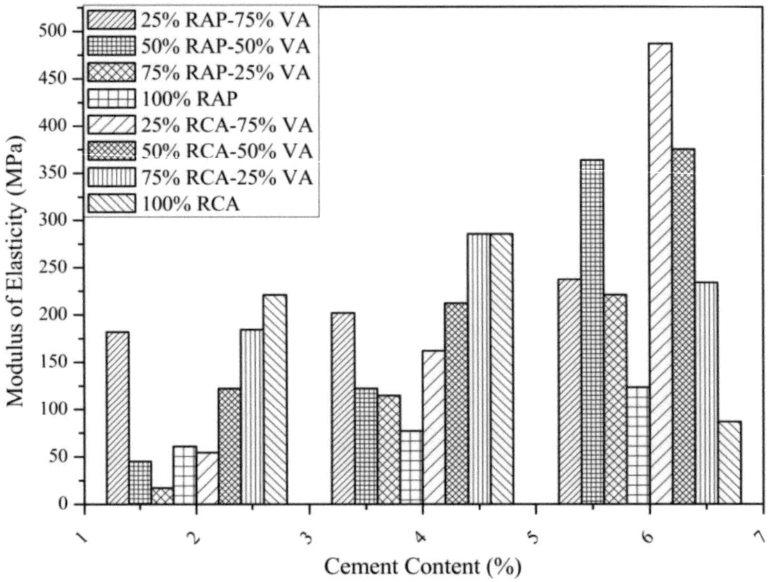

Figure 5.
Modulus of elasticity of RAP VA and RCAVA mixes at 7 days curing period.

RAP did not have a contribution to strength development [24]. A linear relationship is noticed between the UCS and the cement content irrespective of the type of mix. Out of which 25% RAP-75% VA and 50% RCA - 50% VA blends show a rapid gain in strength. This is due to the existence of the unhydrated mortar in RCA, less asphalt coated surface area, and presence of high-quality aggregates, and better interlocking between the aggregates along with the stabilization.

Elastic modulus is the ratio of applied stress to corresponding strain within the elastic limit. This property is used to characterize the materials and to analyze the stresses and strains in different pavement layers. With the continuous application of loads, the recoverable character of these materials will be declined, and the plastic deformation is accumulated. This repeated application of load property is measured using Resilient Modulus (M_R) (**Figure 5**).

The elastic modulus of RAP/VA and RCA/VA blends at different cement contents after 7 days of curing period is calculated. The elastic modulus of RCA blends ranges from 11.95 MPa at 100% RCA with 0% cement content to 486.96 MPa at 25% RCA at 6% cement content. In contrast, the elastic modulus of RAP blended mixes ranges from 60.99 MPa at 100% RAP at 2% cement content to 363.78 MPa at 50% RAP at 6% cement content.

There is a linear increase in the elastic modulus with an increase in the cement content in each mix. However, 100% RCA and 75% RCA-25% VA at 6% cement content there is a tremendous decrease. This is due to the overdosage of the cement to the mix in addition to the existing residual mortar surrounding the aggregates, which makes the material more brittle.

Blending with VA improved the modulus of all RAP mixes as the elastic modulus increased with the increase in the percentage of VA in the combination. The scenario is completely reverse in the case of RCA blends where the elastic modulus increases with the increase in the RCA. This trend is observed up to a smaller dosage of cement contents that is 4%. Whereas at 6% cement, the scenario is completely different for RCA mixes, further addition of the RCA to the mix lowers the modulus values. This clearly shows the effect of blending, in addition to the cement content, equally impacts the overall performance of the mix.

6. Indirect tensile strength test (ITS)

The tensile strength of the cement-treated bases is important as cement-treated materials generally weak in tensile. The developed tensile cracks extend to the top of the pavement layers and weaken the pavement structure makes it susceptible to moisture. Generally, the tensile strain at the bottom of the bituminous layer is considered for analysis which also represents the top of the cement-treated base. The higher the tensile strength represents is more resistance to the tensile stresses that cause in the base. To determine the tensile strength characteristics, the ITS test is carried out on RAP, and RCA blends at different cement contents curing for 7 days. The samples are compacted at the obtained OMC to reach the maximum density with dimensions of the internal diameter of 101.60 mm and 63.5 ± 2.5 mm in height and then extracted after 24 hours followed by curing for 7 days. Then cured samples are tested for ITS as per ASTM D6931 at a loading rate of 50.8 mm per minute, and the failure load is noted. The Indirect Tensile Strength is determined by using the following formula:

$$S_T = \frac{2000P}{\pi D t} \tag{1}$$

Here, S_T is the Indirect Tensile Strength in N/mm^2, D is the Diameter of the Specimen in mm, t is the thickness of the specimen in mm, and P is the Ultimate Failure Load in kN (**Figure 6**).

From **Figure 6**, it is observed that ITS value decreases as the RAP content increases with constant cement; this is due to weak bonding between the RAP and conventional aggregates. 25% RAP with 6% cement shows more ITS value and 50% RCA with 6% cement content have higher ITS in case of RCA blends. The ITS values

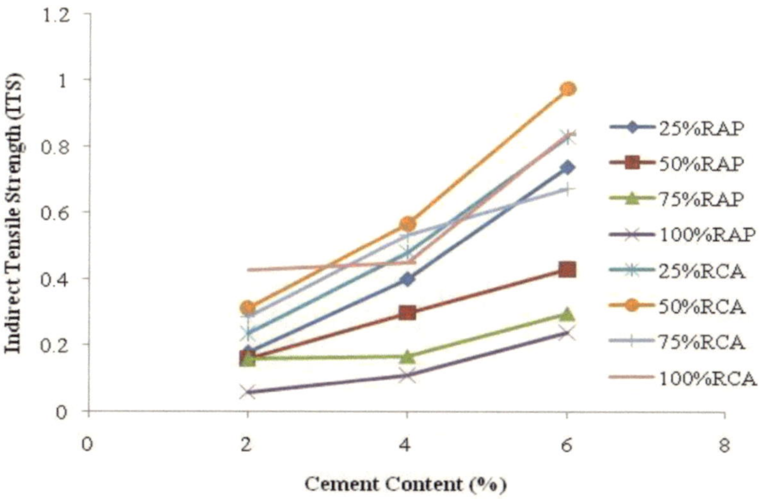

Figure 6.
Indirect tensile strengths of RAP-VA and RCA-VA blends of varying cement percentages [14].

Country/code	ITS (MPa)	
Italy [25]	0.32-0.60 (gyratory compaction) > 0.25 (proctor compaction)	
South Africa [25]	>0.25 for cement 1.5-3%	>0.20 for cement 3-5%

Table 5.
ITS of different countries specifications.

increase with the increase in the amount of cement. The RCA blends show more strength compared with RAP blends as observed in UCS. 50% RCA blends followed by 25% of RCA blends show higher strength compared with remaining blends. This is due to the existence of proper interlocking between RCA and VA and the self-cementing behavior of RCA. In contrast, the RAP blended mixes have a weak bond compared with RCA due to the existing bitumen coating. It is observed that at an average the ITS value is 0.2 times that of the UCS value of RAP treated bases and 0.32 times that of UCS value in case of RCA treated bases for a 7-day curing period.

The ITS value of the present study is compared with other country's specifications. Moreover, it is observed that the acceptable UCS is around 0.20 MPa from **Table 5**. All the recycled aggregate blends achieve this value except 100% RAP at 3% cement content. However, RCA blends achieved the required ITS at 2% cement content.

7. Resilient modulus

Resilient Modulus (M_R) is the ratio of deviator stress to the recoverable strain under the application of repeated loading. It is one of the important stiffness parameters and used as input in the pavement design by most of the transportation departments. The samples were prepared according to ASTM D 1632 and ASTM D 6926 using a cylindrical metal specimen with an interval diameter of 101.60 mm and 63.5 ± 2.5 mm in height. The repeated-load indirect tension test is used to determine the resilient modulus of the mixtures according to ASTM D 4123 by applying compressive loads with a waveform at 25°C temperature and 1 Hz for loading frequencies (the recommended load range can be 10 to 20% of the indirect tensile strength). The Poisson's ratio for the calculation of resilient modulus was assumed as 0.2. The resilient modulus is calculated using the following equation.

$$\text{ERT} = \frac{P(\mu + 0.27)}{\Delta H^{*}t} \quad (2)$$

where P is the repeated load, μ is the Poisson's ratio, ΔH is the horizontal deformation, and t is the thickness of the specimen.

Figure 7 shows that M_R value increases with the cement content and decreases with the RAP content. The maximum stiffness values were observed for 25% RAP mixes

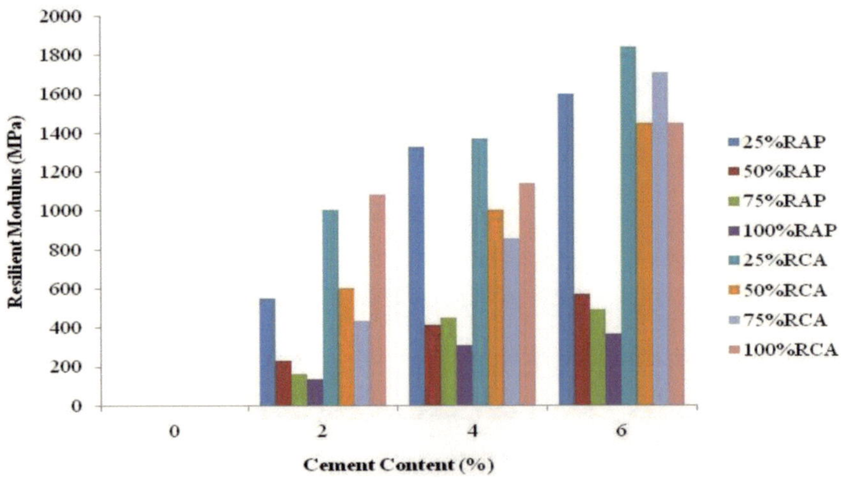

Figure 7.
Resilient Modulus of RAP-VA and RCA-VA mix at 7 days curing period.

and 75% RCA as well. However, there is no appropriate trend that is observed in the RCA blends. However, there is an increase in the stiffness of the mixture independent of the RCA content. As the test is conducted at 7 days of the curing period, more curing periods might be required to gain sufficient stiffness for the cement-treated RAP bases.

8. Applications of the cement-treated recycled bases

The motive of the utilization of the cement-treated recycled bases is to improve the bearing capacity of the base layer with already used aggregates which is a conservative method. It is one of the sustainable construction practices and economical. The selection of the optimum amount of the cement stabilizer is necessary to detrimental overdosage effects like shrinkage. Besides, there is an increase in carbon footprints with a high amount of cement content. The cement-treated recycled bases can be served as a base and subbase layer in the low volume and high-volume roads as well. The Full-depth reclamation, along with cement stabilization, is advantageous when the pavement condition Index is low with poor hydro planning. It will create a strong base layer that can be covered with a thin asphalt layer. Further, the RCA can be used in the base and subbase layers, which is a locally available source. Stabilization of RCA leads good results in decreasing the leachate problems and to improve the mechanical properties.

9. Conclusions

After a thorough investigation of the strength and stiffness properties of the cement-treated recycled materials, the following conclusions are drawn:

- Cement Stabilization of the recycled materials improved the strength and stiffness of the mixes. However, the recycled material content in the mix plays a critical role in the strength development for RAP mixes.

- Maximum strength is achieved at 50% RCA and 25%RAP blended mixes which are measured in terms of ITS and UCS.

- The relation between the ITS and UCS of the cement-treated based is established. On average, the ITS value is 0.2 times that of the UCS value for RAP treated bases and 0.32 times that of UCS value in the case of RCA treated bases at 7 days of curing period.

- Cement stabilization of RCA blends is more effective compared with RAP blends in terms of mechanical properties.

- Based on the experimental results, cement stabilized recycled materials require more curing period to achieve adequate strength and stiffness. All the stabilized bases achieved the target strength at 6% cement content for low volume roads subbases except 100% RAP mix.

Author details

Sarella Chakravarthi, Galipelli Raj Kumar and Sabavath Shankar*
Department of Civil Engineering, National Institute of Technology, Warangal, India

*Address all correspondence to: ss@nitw.ac.in

References

[1] Zhang Y, Liu X, Xu Y, Tang B, Wang Y, Mukiza E. Preparation and characterization of cement-treated road base material utilizing electrolytic manganese residue. Journal of Cleaner Production. 2019; 232:980-92.

[2] Liu J, Yu B, Wang Q. Application of steel slag in cement-treated aggregate base course. Journal of Cleaner Production. 2020:121733.

[3] Arulrajah A, Disfani MM, Haghighi H, Mohammadinia A, Horpibulsuk S. Modulus of rupture evaluation of cement stabilized recycled glass/recycled concrete aggregate blends. Construction and Building Materials. 2015; 84:146-55.

[4] Kasu SR, Manupati K, Muppireddy AR. Investigations on design and durability characteristics of cement-treated reclaimed asphalt for base and subbase layers. Construction and Building Materials. 2020; 252:119102.

[5] Guthrie, W. Spencer, Dane Cooley, and Dennis L. Eggett. Effects of reclaimed asphalt pavement on mechanical properties of base materials. Transportation Research Record. 2005; 1: 44-52.DOI: 10.3141/2005-06

[6] Puppala, Anand J., Laureano R. Hoyos, and Ajay K. Potturi. Resilient moduli response of moderately cement-treated reclaimed asphalt pavement aggregates. Journal of Materials in Civil Engineering. 2011; 23(7): 990-998.DOI: 10.1061/(ASCE)mt.1943-5533.0000268

[7] Murray Reid, J., Khaled E. Hassan, Okan Sirin, and Ramzi A. Taha. Demonstrating the Worth of Recycled Aggregates—A Case Study from Qatar. In Geo-Chicago 2016. p. 534-545. DOI: 10.1061/9780784480137.051

[8] Pérez, Pablo, Francisco Agrela, Rosario Herrador, and Javier Ordoñez. Application of cement-treated recycled materials in the construction of a section of road in Malaga, Spain. Construction and Building Materials. 2013; 44: 593-599. DOI: 10.1016/j.conbuildmat.2013.02.034

[9] Eren, Ş. and Filiz, M. Comparing the conventional soil stabilization methods to the consolid system used as an alternative admixture matter in IspartaDarıdere material. Construction and Building Materials. 2009; 23 (7): 2473-2480.DOI: 10.1016/j.conbuildmat.2009.01.002. Available from: http://www.sciencedirect.com/science/article/pii/S0950061809000038.

[10] Arulrajah, A., Piratheepan, J., Disfani, M. M., & Bo, M. W. Geotechnical and environmental properties of recycled construction and demolition materials in pavement subbase applications. Journal of Materials in Civil Engineering. 2013; 25(8), 1077-1088.DOI: 10.1061/(ASCE)MT.1943-5533.0000652

[11] Arulrajah, Arul, SahanPerera, Yat Choy Wong, SuksunHorpibulsuk, and FarshidMaghool. Stiffness and flexural strength evaluation of cement stabilized PET blends with demolition wastes. Construction and Building Materials. 2020; 239: 117819.DOI: 10.1016/j.conbuildmat.2019.117819

[12] Arshad M. Laboratory investigations on the mechanical properties of cement-treated RAP-natural aggregate blends used in base/subbase layers of pavements. Construction and Building Materials. 2020; 254:119234.

[13] Yan K, Li G, You L, Zhou Y, Wu S. Performance assessments of open-graded cement stabilized macadam containing recycled aggregate. Construction and Building Materials. 2020; 233:117326.

[14] Chakravarthi, S., Anusha Boyina, Arun Kumar Singh, and S. Shankar. Evaluation of cement-treated reclaimed asphalt pavement and recycled concrete pavement bases. International Journal of Pavement Research and Technology. 2019; 12 (6): 581-588.DOI: 10.1007/s42947-019-0069-1

[15] Faysal, M., Mahedi, M., Aramoon, A., Thian, B., Hossain, M. S., Khan, M. A., & Khan, M. S. Determination of the structural coefficient of different combinations of cement-treated/untreated recycled base materials. In Geotechnical and Structural Engineering Congress; 2016. p. 1198-1208. DOI: 10.1061/9780784479742.100

[16] LaHucik, Jeffrey, Scott Schmidt, Erol Tutumluer, and Jeffery Roesler. Cement-treated bases containing reclaimed asphalt pavement, quarry by-products, and fibers. Transportation Research Record, 2580. 2016; (1): 10-17. DOI: 10.3141/2580-02

[17] Behiry, Ahmed Ebrahim Abu El-Maaty. Utilization of cement treated recycled concrete aggregates as base or subbase layer in Egypt. Ain Shams Engineering Journal. 2013; 4 (4): 661-673.DOI: 10.1016/j.asej.2013.02.005

[18] Xuan, D. X., L. J. M. Houben, A. A. A. Molenaar, and Z. H. Shui. Mixture optimization of cement treated demolition waste with recycled masonry and concrete. Materials and structures 45. 2012; 1-2: 143-151.DOI: 10.1617/s11527-011-9756-3

[19] Taha, Ramzi, Ali Al-Harthy, Khalid Al-Shamsi, and Muamer Al-Zubeidi. "Cement stabilization of reclaimed asphalt pavement aggregate for road bases and subbases." Journal of materials in civil engineering 14, no. 3 (2002): 239-245.DOI: 10.1061/(asce)0899-1561(2002)14:3(239)

[20] Ministry of Road Transport and Highways, Specifications for Road and Bridgeworks, Fifth Revision, Ministry of Road Transport and Highways, New Delhi, 2013.

[21] Guthrie WS, Brown AV, Eggett DL. Cement stabilization of aggregate base material blended with reclaimed asphalt pavement. Transportation Research Record. 2007; 2026(1): 47-53. DOI: 10.3141/2026-06

[22] MORD. Ministry of Rural Development: Specifications for Rural Roads. New Delhi: The Indian Roads Congress, 2014.

[23] Poon CS, Chan D. Feasible use of recycled concrete aggregates and crushed clay brick as unbound road sub-base. Construction and building materials. 2006; 20(8): 578-585. DOI: 10.1016/j.conbuildmat.2005.01.045

[24] Yuan D, Nazarian S, Hoyos LR, Puppala AJ. Evaluation and mix design of cement-treated base materials with a high content of reclaimed asphalt pavement. Transportation Research Record. 2011; 2212 (1):110-119.DOI: 10.3141/2212-12

[25] Autelitano, F., & Giuliani, F. Electric arc furnace slags in cement-treated materials for road construction: Mechanical and durability properties. Construction and Building Materials. 2016; 113: 280-289.DOI: 10.1016/j.conbuildmat.2016.03.054

Chapter 12

Cementitious Grouts Containing Irradiated Waste Polyethylene Terephthalate

Muhammad Imran Khan, Muslich Hartadi Sutanto, Madzlan Bin Napiah and Salah E. Zoorob

Abstract

This chapter describes a review of the design and formulation of various cementitious grouts for semi-flexible pavement surfaces. Additionally, the authors also conducted extensive experimental work on the possibility of using a most effective and innovative way of recycling waste polyethylene terephthalate (PET) by exposing to gamma radiation and using as a replacement of Ordinary portland cement in the formulation of cement grouts for semi-flexible pavement surfaces. In the current study, cement in the grouts was replaced with PET (regular and irradiated), fly ash and silica fume and was evaluated for flowability and strength properties. The study concludes that normal PET causes a significant reduction in compressive strength, however, some of the strength is restored when irradiated PET was used. The recycling of waste PET, as a cement replacement in the cementitious grouts for semi-flexible pavement surfaces, with the irradiation process can be doubled as compared to utilizing normal/regular PET.

Keywords: cementitious grout, irradiated waste polyethylene terephthalate, fly ash, silica fume, compressive strength

1. Introduction

Generally, pavements are classified into two types: flexible pavements and rigid pavements. Conventional flexible pavements are constructed from bituminous materials and are widely used as a highway, expressway/freeway, and airport pavements due to their satisfactory performance against distresses and better riding quality, good serviceability, high skid resistance, low cost and easy maintenance [1–3]. However, due to recent exponential increase in traffic load and extreme adverse environmental conditions, the flexible pavements are exposed to many distresses (such as rutting, cracking, corrugation, shoving, stripping etc.) which can badly affect its service life and performance [4, 5]. On the other hand, rigid pavements are constructed from cement concrete with or without reinforcement. Rigid pavements have better durability, high compressive strength but have some disadvantages such as; provision of joints, rough-riding quality, slow setting time, high susceptibility to thermal stresses, high initial cost and maintenance efforts, cannot simply be ignored [6–8]. Taking into consideration the disadvantages of both flexible and rigid pavements there was a need for an alternative pavement design that can combine the positive attributes of conventional flexible and rigid pavements with a view of the improvement in serviceability and performance. One such alternative is the design of semi-flexible pavement surfaces (SFPS) [9–13].

2. Semi-flexible pavement surfaces

Semi-flexible pavement surface (SFPS) material (also known as grouted macadam) is designed as an open-graded asphalt mixture (OGAM) with 25–35% air voids and then cement grout is infused into the voids [2, 7, 14–17]. However, few researchers advised the range of air voids to be 20–35% [18, 19]. It is a well-known fact that the major demerits of rigid pavements are extended construction time and provision of joints to allow thermal expansion as compared to flexible pavements. Therefore, the semi-flexible pavements are gaining popularity from last few decades and are devised as a jointless pavement with reduced time of construction, easy to repair and construct as well as offers good serviceability and riding quality as compared to rigid pavements [18, 20, 21]. This new type of pavements possesses both flexibility and rigidity which depends on characteristics of the bituminous mixture and cement grout, respectively. Moreover, SFPS provides superior resistance to rutting and are not susceptible to permanent deformation [18]. Semi-flexible pavements have been constructed in various countries as a highway intersection, bus lanes in urban areas, tunnels roads, the pavement in industrial area for the movement of heavy machinery, airport taxiways and aprons as fuel and spillage resistance [15, 16, 21–27].

3. History of the construction of semi-flexible pavement

The first construction of SFPS was carried out in the 1950s, in France and was given the name as Salviacim. It was proposed to provide resistance against oil and fuel attack on the surface and was constructed as a protective layer over conventional asphalt concrete pavement [28]. However, further development of Salviacim was processed by Lefebvre Enterprises, a French construction company, as a replacement to rigid pavements to provide cost-effective construction [29, 30]. Later, this type of pavement construction spread in different countries including South Africa, Saudi Arabia, Australia, UK and Japan as a heavy-duty surface construction [31]. Some other brand names also become popular in Europe based on the type of designed materials such as Hardicrete Heavy-duty surfacing, Densiphalt (or confalt) and Worthycim Heavy-duty surfacing [32–35]. In the USA the semi-flexible pavement is called as Resin Modified Pavement (RMP) and in Japan, it was given name as Rut-proof Pavement (RP-Pavement) [20, 29]. Moreover, the construction of SFPS spread throughout Europe, Various states of Africa, North America, South Pacific and Far East [36].

4. Composition of semi-flexible pavement surfaces

The construction of semi-flexible pavement surfaces (SFPS) is performed in two stages. In the first stage, the open-graded asphalt mixture (OGAM) is prepared, laid, and compacted to achieve the target air voids of 25–35%. The compaction is accomplished by a light application of roller compactor without vibration to avoid disintegration of course aggregates and tracks in the mixture. In the second stage, the cementitious grouts are prepared and poured on the surface of OGAM and allowed to infiltrate into the voids. The construction is normally carried out in two consecutive days, the reason is to allow the OGAM to cool down before applying the cement grouts. Rubber scraper can be used to spread the grout on the surface [18, 20]. The schematic representation of this process is demonstrated in **Figure 1**. Hence the major constitutes of semi-flexible surfaces are a selection of open-graded asphalt mixture and formulation of cement grouts.

4.1 Selection of open graded asphalt mixture

The major constituent in constructing semi-flexible pavement surfaces are the aggregates. Consequently, it needs careful attention while selecting gradation of aggregate for mix designing of semi-flexible mixtures. Typically, semi-flexible pavement surfaces comprise of open-graded asphalt skeleton with air voids ranging from 25 to 35% above which highly flowable cementitious grouts are spread and allowed to penetrate [35, 37, 38]. The key parameters for the selection and design of OGAM for semi-flexible pavements are the consideration of air voids and binder drainage. The air voids of final compacted OGAM shall be in the range of 25 to 35%. In case if air voids are less than 25%, the mixture may not have interconnected voids and would be difficult for the cement grout to penetrate through the depth of OGAM [39]. As a result, adequate strength and homogeneous properties will not be attained to withstand traffic load and environmental stresses as shown in **Figure 2** [18]. On the other hand, if air voids are more than 35%, a large amount of cement grout will be required, and the final mix will more be likely a rigid pavement and may lead to fatigue failure. Similarly, binder drainage (or draindown) of bitumen in the final mixture shall not be more than 0.3% as per AASHTO T 305 or ASTM D6390 standards [40, 41].

Therefore, few studies have been conducted for the selection of suitable aggregate gradation that fulfills the requirement of air voids and binder drainage. Recently, a detailed study has been conducted by Saboo N. et al., (2019) on

Figure 1.
Schematic representation of the process of preparing semi-flexible surfaces.

Figure 2.
Disconnectivity of internal voids due to low air-voids [18].

evaluating different types of porous aggregate gradations for semi-flexible pavement surfaces. The hierarchical ranking approach was used to reject/select and finally to choose the most desirable gradation system [26]. In this study, seven different porous asphalt mix skeletons were prepared with bitumen contents ranging from 2–5%. The gradations were selected from the previous research that has already been used in various studies and is presented in **Figure 3**.

During the initial elimination process, draindown test, air voids in the compacted mix, voids in the coarse aggregate (VCA), permeability and cantabro loss tests were used. The selected mixtures were further evaluated for Indirect Tensile test (ITS) to rank the gradation system accordingly. It was finally suggested three mixes including BSI with 4% bitumen, Densiphalt-12 with 4% and Densiphalt-12 with 4.5% bitumen were chosen as most desirable porous asphalt mix skeleton to be considered for semi-flexible pavement surfaces as shown in **Table 1** [26].

Similarly, another study was conducted by Hou et al., on the random selection of 22 aggregate gradations by varying percentages of coarse aggregates, fine aggregates, and fillers. The effect of grouting on these gradations was evaluated by volumetric analysis, cantabro and binder drainage test as well as wheel tracking and bending tests [42]. It was revealed that grouting ability of OGAM is not only influenced by initial air voids but also on void interconnectivity, morphological characteristics, and size of pores. The comparison of air voids before and after grouting is demonstrated in **Figure 4** and can be seen that grouting ability is not

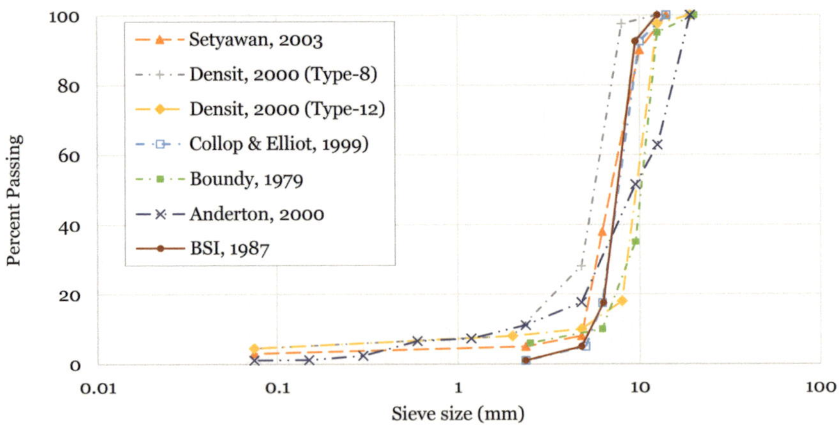

Figure 3.
Porous asphalt gradation curves.

Gradation	Draindown (%) <0.30	Air voids (%) >25%	VCA ratio < 1	Permeability (m/day) >100 m/day	Cantabro loss (%) <50%	ITS(kPa)
BSI with 4% bitumen	0.30	33.05	0.95	362.90	33.20	113.35
Densiphalt-12 with 4%	0.19	33.56	0.93	332.90	34.91	114.42
Densiphalt-12 with 4.5%	0.30	32.08	0.91	247.69	34.91	136.80

Table 1.
Final selection criteria for porous aggregate gradation.

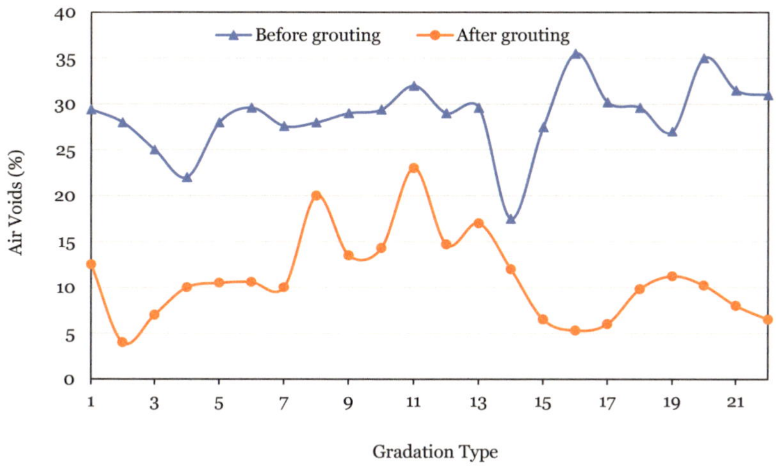

Figure 4.
Air voids before and after grouting [42].

only influenced by initial air voids. The grouting ability can also be enhanced by increasing the percentage of coarse aggregates and reducing binder content (3.8% in this study was used) which can improve the interconnectivity of voids.

4.2 Cement grouts

Cement grout or cement slurry is an essential part of semi-flexible pavement surfaces, which contributes to the rigidity of pavement. The design and formulation of grouts for semi-flexible pavement surfaces differ from traditional cement past and cement mortars used in the concrete industry. The constituents of the grouts are formulated in such a way that can easily be penetrated the voids of OGAM. Additionally, the grouts should have enough strength to withstand the stresses induced due to traffic and the environment. Therefore, the flowability (fluidity), workability, and strength of cement grouts play a key role in the performance of semi-flexible pavement surfaces. In a scenario, if the fluidity of cement grout in insufficient it would not fill properly the air voids of OGAM and as a result, the designed SFP mixtures would not provide sufficient strength and durability. The fluidity of grouts is usually determined by flow-cone, however, the flow-out time depends on the geometry of the cone. Ordinary Portland cement (OPC), water (different ratios), and/or sand, fillers, and other supplementary cementing materials (such as fly ash and silica fume) are normally used to formulate the compositions of cement grouts. Furthermore, superplasticizer can be used to improve the fluidity at a relatively low w/c ratio. Various studies have been conducted on designing the constituents of cement grouts for semi-flexible pavements surfaces and are listed in the following section.

5. Design parameters of cement grouts for SFPS

5.1 Flowability of the cement grouts

One of the primary considerations to formulate the compositions of cementitious grouts for semi-flexible pavement surfaces is the flow or fluidity of grouts. The grouts are required to be sufficiently flowable that can easily penetrate the voids of the porous asphalt skeleton. The flowability/fluidity of grouts highly depends on w/c ratio, superplasticizer as well as other additives and supplementary cementing materials. Traditionally, the flowability/fluidity is measured by flow-cone apparatus in which the desired quantity of grout is poured and the flow-out time of grout from the flow-cone is measured in seconds. Higher the time of flow, lower is the fluidity and hence less workability. On the other hand, a lower time taken by grout to flow-out indicates high flowability and higher workability.

However, the standard requirement of flow-out time of cement grout depends on the geometry and size of the flow-cone, as three different flow-cones are being frequently used in literature. Which are; (a) Malaysian flow-cone, (b) Marsh flow-cone and (c) ASTM flow-cone and their schematic representation are shown in **Figure 5(a-c)**. According to the Malaysian standards, the flow-out time of 1000 mL of grout using Malaysian flow-cone (**Figure 5(a)**) shall be in the range of 11 sec to 16 sec [43, 44]. The cement grouts with the flow of 11–16 sec will have sufficient fluidity and suitable for grouting SFP surfaces. Similarly, the quantity of grout required for Marsh flow-cone (**Figure 5(c)**) is also 1000 mL, however, due to change in geometry the time of flow-out shall be in the range of 8 to 10 sec [38]. Nevertheless, the flow-cone used in ASTM C939 have different geometry as shown in **Figure 5(b)** and hence the requirement of flow-out time. According to the standard, recommended fluidity of cement grout is 10 to 14 sec while allowing a grout quantity of 1725 mL to flow-off the funnel.

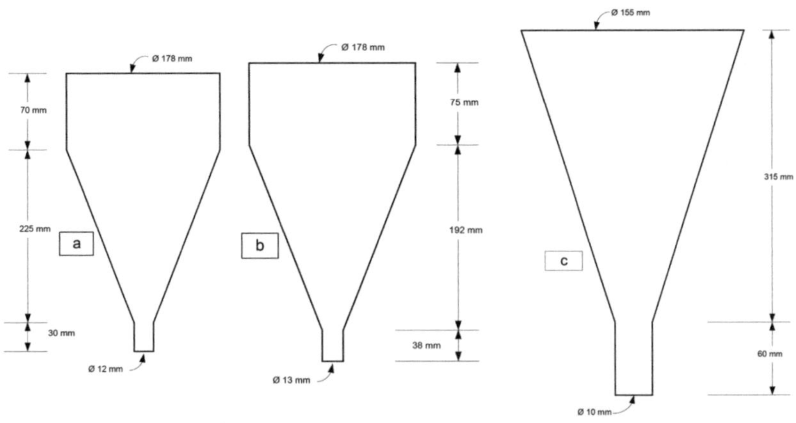

Figure 5.
Flow cones used in literature to measure the fluidity of grouts: (a) Malaysian flow-cone, (b) ASTM flow-cone and (c) marsh flow-cone (dimensions not-to-scale).

Various studies have been conducted on evaluating the fluidity of cement grout by varying water content as well as inclusion superplasticizer, other additives and supplementary cementing materials. However, increasing w/c ratio causes a reduction in strength properties of grout and therefore it shall be carefully selected. To overcome the issues related to high w/c ratio, superplasticizer plays a key role in improving the fluidity of grouts at relatively low w/c ratio. It is always recommended to use superplasticizer in designing cement grouts for SFP surfaces at low w/c ratio in order to achieve the desired strength properties at high flowability of grouts [15, 45]. A study concluded that increasing w/c ratio causes a significant increase in fluidity with a considerable reduction in compressive strength. However, the addition of superplasticizer up to 1% (by weight of cement) causes improvement in fluidity as well as compressive strength. Moreover, increasing superplasticizer beyond 1.5% can cause a bleeding problem in grouts [46].

The incorporation of high w/c ratio and plasticizing action of styrene-butadiene admixtures produces highly flowable cement grout however with low strength properties and gives weaker grouts [15]. On the other hand, stronger grouts with desired fluidity can be achieved by the addition of milled glass and polycarboxylate superplasticizer at relatively low w/c ratio [15]. The highest fluidity presented by the weaker grout group could be explained by their higher water/cement ratio and the styrene-butadiene admixture plasticizing action. In the stronger grouts group, increasing the content of milled glass significantly reduces the flow time values. Incorporating milled glass particles with their impermeable and low specific surface areas indicates that at the same water/cement ratio water is available to fluidize the grouts.

Similarly, in a study different w/c ratio (0.48–0.63), fly ash content (0–20%) and mineral powder (0–20%) were used to design cement paste and to analyze the fluidity of grouts using flow-cone shown in **Figure 5(b)**. Increasing w/c ratio causes a significant increase in the fluidity of cement paste. Similarly, the increase in fluidity was also observed with increase in the dose of fly ash. However, the addition of mineral powder initially causes a reduction in fluidity and then increased was witnessed. Based on fluidity, drying shrinkage and strength properties, 0.56–0.58 w/c ratio, 10% fly ash and 10% mineral powder were recommended as a suitable combination of grouts for SFP mixtures [47].

5.2 Strength properties of grouts

The second most important parameter for selection of cement grout for SFP surfaces is the compressive strength of grouts. The compressive strength largely depends on w/c ratio and other supplementary cementing materials and additives. Various materials have been used so far including, fly ash, silica fume, ground granulated blast furnace slag (GGBS), gypsum and mineral powders for designing cement grouts suitable for SFP surfaces requirements. A study has shown that increase in w/c ratio from 0.48 to 0.63 causes a significant reduction in 7 and 28-days compressive strength, however, flexural strength was slightly reduced [47]. Therefore, there is need to introduce the addition of other admixtures and supplementary cementing materials, such as superplasticizer, fly ash, silica fume which can contribute in improving the strength properties at relatively low w/c ratio while achieving the desired flowability.

The combination of silica fume additionally with superplasticizer can produce grouts with sufficient fluidity and strength properties that can be recommended for semi-flexible pavement surfaces. Hence, silica fume with 5% replacement of OPC and 2.0% polycarboxylate based superplasticizer at 0.30 w/c ratio gives fluidity of 15 sec (as determined by Malaysian flow-cone) and 28-days compressive strength of 92.5 MPa [48]. Therefore, this grout can be recommended to produce high-strength cement grouts for semi-flexible pavement surfaces that can be used in heavy-loaded pavements.

6. Formulation of cement grouts for SFPS

It is necessary to optimize the constituents of cement grouts that can be used as a grouting material for semi-flexible surfaces. The constituents are but not limited to water, cement, superplasticizer, sand, and/or other supplementary cementing materials (SCM). The optimization is normally evaluated based on fluidity and compressive strength (at different curing period). The Authors also conducted a related study to optimize the constituents of cement grout using a statistical tool known as response surface methodology (RSM). In this study, w/c ratio (0.25 to 0.45) and Polycarboxylate-ether type superplasticizer (0 to 2%) were selected as factors (dependent variables) whereas flow-value and compressive strength (1-day, 7-days, and 28-days) were considered as responses (dependent) variables to conduct statistical analysis and optimization in RSM. The ANOVA and multi-objective optimization techniques were utilized to formulate the optimum combination of w/c ratio and dosage of superplasticizer [46].

A study conducted by [49] utilizing latex modified cement mortar to investigate the performance SFPS. The composition of cement grouts was selected as; 0.70 w/c ratio, 20% sand, 10% limestone filler, and latex with 0, 1.2, and 2.4%. The results indicate that an increasing percentage of latex improved the compressive and flexural strength, however, a negative effect on fluidity. Moreover, SFP mixtures with latex-modified grout showed better moisture resistivity, rutting resistance, and fatigue life as compared to conventional asphalt mixture [49]. In another study, three different interface optimizers (silane coupling agent, carboxyl styrene-butadiene latex, and cationic emulsified asphalt) were used with various percentages in cement grouts to improve the interfacial connection between the cement grout and asphalt in SFP mixtures. The results indicate that interface optimizers despite causing a reduction in strength properties of grout, greatly improves the interfacial connection of cement grout with asphalt as observed from microstructural analysis of mixtures as shown in **Figure 6** [50]. These interface optimizers also show a positive effect in terms of crack resistance at low temperature, stability at high temperature, and moisture stability, particularly by cationic emulsified asphalt.

A similar study was conducted to evaluate the adhesion of cement grout with porous asphalt mixture and the mechanical properties of SFP mixtures. In this study a modified cement grout containing; cement, w/c ratio (0.40), silica fume (10%), superplasticizer (2%), aluminum powder (0.04%), viscosity modifying agent (0.2%), and asphalt emulsion (20%, 40% and 60%) were used to produce SFP mixtures. The results indicate that asphalt emulsion modified grouts have better adhesion with porous asphalt mixture as compared to control grouts. This phenomenon may overcome fatigue problems in semi-flexible pavements. The overall performance of semi-flexible pavement mixtures using asphalt emulsion modified grouts showed better performance in terms of rutting resistance, moisture damage resistance, and low-temperature crack resistance as compared to control grout and conventional HMA [1].

An analytical approach using mathematical programming was used to optimize the constituents of cement grouts. Various combinations of grouts were prepared using w/c (0.40, 0.50, 0.60), naphthalene based superplasticizer (0%, 2%, 4%, 6%),

Figure 6.
Microstructure of cement-asphalt interface modified with a silane coupling agent, carboxyl styrene-butadiene latex, and cationic emulsified asphalt [50].

sand with 23% by weight of cement (size <2.36 mm, <1.18 mm, <0.6 mm) to optimize the composition of grouts by evaluating flow and compressive strength properties. The results indicate that flow was significantly improved by increasing the dose of naphthalene superplasticizer from 0 to 6% and the w/c ratio from 0.40 to 0.60. The optimal combination of grouts obtained using mathematical programming was 0.54 w/c ratio, 2% superplasticizer, and 1.66 mm sand size. The flow time and compressive strength at this combination are 14 sec and 24.25 MPa respectively. Similarly, another group of grouts was prepared using polycarboxylate superplasticizer. The compositions were w/c ratio (0.40, 0.45, 0.50), polycarboxylate superplasticizer (0%, 0.5%, 1.0%, 1.5%, 2.0%), sand with 23% by weight of cement (size <2.36 mm, <1.18 mm, <0.6 mm). Interestingly, the compressive strength of grouts increased with an increasing dose of SP from 0 to 1.0% and then starts decreasing again. Moreover, the flow time was significantly decreased with increasing SP and w/c ratio. The optimal doses of ingredients using the analytical approach were 0.48 w/c ratio, 1.0% SP and 0.6 mm sand particle size with flow time of 14 sec and compressive strength of 23.46 MPa [27].

Latex powder (0%, 1.2% and 2.4% by weight of cement) was used to modify cement grout with 0.72 w/c ratio, 20% sand, and 10% filler and to use for semi-flexible pavement mixtures. The addition of latex powder showed improved compressive and flexural strength of cement grout, while causes reduction in fluidity. The latex powder with 1.2% was considered as optimum based on strength and fluidity and was used to produce SFP mixtures. SFP mixtures with 1.2% latex-modified grout showed better resistance to high-temperature performance while higher moisture susceptibility, while weaker fatigue resistance and low-temperature performance. However, the moisture damage and low-temperature behavior of SFP mixtures were still in good agreement [51]. Similarly, cement grout was formulated using 0.63 w/c ratio, 10% mineral filler, 20% sand and 0–12% Carboxyl Latex. The substation of carboxyl latex to grout causes a reduction in compressive strength both at 7-days and 28-days curing.

However, flexural strength was increased with increasing dose of carboxyl latex. Hence, 8% of carboxyl latex was considered as optimum dosage. The performance of SFP mixtures filled with carboxyl latex-modified grout was improved in terms of rutting resistance, low-temperature crack resistance, moisture damage and fatigue resistance [52]. Moreover, various studies have been conducted on the selection of compositions of cement grouts for semi-flexible pavement surfaces as described in **Table 2**.

Composition of grouts	Strength Properties of grouts	Concluding remarks	References
w/c ratio = 0.50, fly-ash = 23%, superplasticizer = 2%	Not determined	SFP mixtures demonstrate higher resistance against rutting and moisture damage as compared to HMA mixtures. SFP mixtures showed lower fatigue life as compared to HMA mixtures. SFP mixtures showed better thermal cracking resistance at low temperatures than HMA mixture using fracture work property.	[53]
w/c ratio = 0.31 Cement grouting was JGM-301 (factory produced)	Compressive Strength: 3-days = 34.5 MPa 28-dyas = 42.3 MPa	It was recommended that semi-flexible pavement surfaces with this grout can be used for heavily loaded pavements.	[19, 54]
Water: cement: sand: filler = 720:1000:497:249 Means: w/c ratio = 0.72, sand =20%, filler = 10% Latex = 0, 1.2%, 2.4%	Compressive Strength: 7-days = 21.6 MPa to 24.1 MPa Flexural Strength: 7-days = 1.65 MPa to 3.54 MPa	Latex in cement mortar has a positive effect on compressive and flexural strength while negative impact on fluidity. SFP mixtures show better moisture resistance, high rutting resistance, and better fatigue life as compared to dense asphalt mixtures. However, SFP mixtures showed poor performance in terms of brittle cracking resistance as compared to dense asphalt mixtures.	[49]

Composition of grouts	Strength Properties of grouts	Concluding remarks	References
Three interface optimizers were used 1. Control grout (OPC) 2. Silane coupling agent (0.25%, 0.50%, 0.75%) 3. Carboxyl styrene-butadiene latex (5%, 10%, 15%) 4. Cationic emulsified asphalt (5%, 10%, 15%) w/c ration = not known	7-days compressive Strength: 1. 42.98 MPa for OPC 2. 34.65 MPa, 36.85 MPa, 33.61 MPa for Silane coupling agent 3. 25.70 MPa, 29.70 MPa, 26.15 MPa for Carboxyl styrene-butadiene latex 4. 37.79 MPa, 33.82 MPa, 31.53 MPa for Cationic emulsified asphalt	The interface optimizers improve the drying shrinkage resistance but cause a reduction in strength properties. The interfacial connection of cationic emulsified asphalt modified grout with bitumen was proved to be best whereas worst for silane coupling agent. These interface optimizers also deliver satisfactory results in terms of crack resistance at low temperature, stability at high temperature, and moisture stability.	[50]
w/c ratio = 0.40 silica fume = 10% SP = 2% aluminite powder (AP) = 0.04% viscosity modifying agent (VMA) = > VMA/C = 0.2% Asphalt Emulsion (AE)/cement binder AE/C = 20%, 40%, 60%	7-days compressive Strength: CP = 26 MPa CAEP20% = 13 MPa CAEP40% = 11.3 MPa CAEP60% = 10.5 MPa 28-days compressive Strength: CP = 28.5 MPa CAEP20% = 19 MPa CAEP40% = 17.8 MPa CAEP60% = 17 MPa	The compressive strength of grout reduces with the addition of asphalt emulsion. The asphalt emulsion improves the adhesion between cement grout and porous asphalt mixture. SFP with asphalt emulsion-based grouts has better rutting resistance as compared to conventional HMA. However, the rutting resistance decrease with increasing percentages of asphalt emulsion in the grouts. The overall performance of SFP mixtures based on asphalt emulsion modified grouts have better performance in terms of rutting resistance, moisture damage resistance, and low-temperature crack resistance as compared to control grout and conventional HMA.	[1]

Composition of grouts	Strength Properties of grouts	Concluding remarks	References
Naphthalene based superplasticizer = 0, 2%, 4%, 6% w/c ratio = 0.4, 0.5, 0.6 Sand = 23% by weight of cement	7-days compressive Strength: 10.22 MPa to 31.65 MPa (for sand size <0.6 mm) 13.15 MPa to 33.75 MPA ((for sand size <1.18 mm) 16.48 MPa to 36.69 MPA ((for sand size <2.36 mm)	An analytical approach was used for optimization using mathematical programming. The optimal composition of grout: w/c ratio = 0.54, SP = 2% with compressive strength of grout = 24.25 MPa	[27]
Polycarboxylate superplasticizer = 0, 0.5%, 1%, 1.5%, 2% w/c ratio = 0.4, 0.45, 0.50 Sand = 23% by weight of cement	7-days compressive Strength: 13.15 MPa to 29.10 MPa (for sand size <0.6 mm) 16.38 MPa to 30.39 MPA ((for sand size <1.18 mm) 18.45 MPa to 31.20 MPA ((for sand size <2.36 mm)	An analytical approach was used for optimization using mathematical programming. The optimal composition of grout: w/c ratio = 0.48, SP = 1% with compressive strength of grout = 23.46 MPa	[27]
w/c ratio = 0.72, sand =20%, filler = 10% Latex powder (LP) = 0, 1.2%, 2.4%	7-days Compressive Strength: 23.5 MPA (0% LP), 24.1 MPa (1.2% LP), 21.6 MPa (2.4% LP) 7-days Flexural Strength: 1.65 MPa (0% LP), 2.81 MPa (1.2% LP), 3.54 MPa (2.4% LP)	Fluidity decreases with increasing latex powder. The cement grout with 1.2% latex powder has higher compressive strength as compared to control and 2.4% latex powder. However, the flexural strength of grouts increased with increasing latex powder. SFP mixtures with latex-modified grout showed better resistance to high-temperature performance and moisture susceptibility, while weaker fatigue resistance and low-temperature performance.	[51]

Composition of grouts	Strength Properties of grouts	Concluding remarks	References
w/c ratio = 0.63, mineral filler = 10% and sand = 20% Carboxyl Latex = 0–12%	7-days Compressive Strength: 20.76 MPa to 11.20 MPa 28-days Compressive Strength: 31.43 MPa to 18.56 MPa 7-days Flexural Strength: 1. MPa to 4.95 MPa 28-days Flexural Strength: 5.74 MPa to 6.49 MPA	The substation of carboxyl latex to grout causes a reduction in compressive strength both at 7-days and 28-days curing. However, flexural strength was increased with increasing dose of carboxyl latex. 8% carboxyl latex was considered as optimum dosage. The performance of SFP mixtures filled with carboxyl latex-modified grout was improved in terms of rutting resistance, low-temperature crack resistance, moisture damage and fatigue resistance.	[52]
w/c ratio = 0.60 cement: sand = 1: 0.5	3-days compressive strength = 10.4 MPa 28-dyas compressive strength = 30.7 MPa 3-days Flexural Strength = 3.3 MPa 28-days Flexural Strength = 7.3 MPa	The semi-flexible mixtures show better high-temperature performance while maintaining its flexibility. It also demonstrates better results of low-temperature cracking resistance as compared to conventional HMA. However, the split tensile strength was lowered. The SFP mixtures showed significant improvement in moisture resistance and present great advantages over conventional HMA.	[55]

Composition of grouts	Strength Properties of grouts	Concluding remarks	References
Cement paste: w/c ratio = 0.48, 0.53, 0.58, 0.63 fly ash = 0%, 10%, 20% mineral powder (ground granulated blast furnace slag) = 0%, 10%, 20% **Cement mortar:** w/c ratio = 0.55, 0.60, 0.65 fly ash = 0%, 10%, 20% mineral powder (ground granulated blast furnace slag) = 0%, 10%, 20% Sand = 10%, 15%, 20%	**Optimal Combinations:** Cement paste: 0.56–0.58 w/c ratio, 10% fly ash and 10% mineral powder Cement mortar: 0.61–063 w/c ratio, 10% fly ash and 15% mineral powder	The overall performance of cement paste was observed to be better than cement mortar in terms of fluidity and strength and recommended more suitable for grouting semi-flexible pavement surfaces with the following combinations; w/c ratio = 0.58, fly ash = 10 and mineral powder = 10%	[47]
w/c ratio = 0.30, 0.35, 0.40, 0.45, 0.50 SP (Polycarboxylate) = 0 to 2.5% Silica fume = 0%, 5% and 10% Optimum combination of 0.30 w/c ratio, 2.0% SP and 5% silica fume	1-day compressive strength = 57.5 MPa 28-days compressive strength = 92.5 MPa 7-days Flexural strength = 6.7 MPa 28-days Flexural strength = 9.1 MPa	Grout with 5% replacement of SF and 2.0% SP produces cement grouts with desirable flowability and compressive strength required for SFP surfaces.	[48]

Table 2.
Summary of the various types of cement grouts used in literature.

7. Irradiated polyethylene terephthalate (PET) based cement grouts

The authors utilized an innovative and effective way of recycling waste PET in cement grouts for semi-flexible pavement surfaces. This was done by exposing waste PET to gamma rays, which, in fact, improve the crystallinity and chain-scission properties of PET. The waste PET was obtained from Plastic Recycling Factory, Ipoh, Malaysia. The initial particle size was in the range of 0.075 mm to 0.85 mm. After exposing to gamma rays, the PET was further sieved on 0.015 mm sieve to obtain a fine powder.

Utilizing regular PET in cement grouts causes a significant reduction in strength properties, however, some of the lost strength can be recovered back while using irradiated PET instead of regular PET. Similarly, silica fume and fly ash were also added for the purpose to achieve high strength cement grouts. The elemental and chemical composition of cement (OPC), silica fume and fly ash are given in **Table 3**.Initially, cement grouts were produced by varying water-cement (w/c) ratio (0.25 to 0.45) and superplasticizer (0–2%) and were evaluated for flowability using Malaysian flow cone and compressive strength at 1-day, 7-days and 28-days curing. Based on flowability in the range of 11 sec to 16 sec, maximizing compressive strength and taking into account the bleeding problem, w/c ratio of 0.35 and superplasticizer of 1% were selected. Further grouts were produced by replacing OPC with regular PET, Irradiated PET, fly ash and silica fume. The compositions are demonstrated in **Table 4**.

Material	SiO_2	Al_2O_3	Fe_2O_3	CaO	MgO	K_2O	SO_3	TiO_2	P_2O_5	Na_2O	Others
Cement	22.65	4.63	2.34	61.72	4.23	1.14	2.24	0.20	0.12	0.11	0.62
Silica Fume	92.5	0.92	0.8	0.93	1.6	0.5	0.82	—	—	0.2	1.73
Fly Ash	36.4	13.72	18.24	19	3.26	2.2	2.5	1.45	1.2	1.73	0.3

Table 3.
Elemental and chemical composition of cement, silica fume and fly ash.

Grout symbol	PET % (regular/irradiated)	Fly ash %	Silica fume %	Other constitutes
G1	0	0	0	0.35 w/c ratio 1% SP
G2	2.5	10	—	
G3	5	10	—	
G4	7.5	10	—	
G5	10	10	—	
G6	2.5	—	5	
G7	5	—	5	
G8	7.5	—	5	
G9	10	—	5	

Table 4.
Constitutes of PET based cement grouts.

7.1 Flow properties of PET-based grouts

The flowability/fluidity of cement grouts was determined by measuring the flow-out time of 1-liter grout using Malaysian flow-cone. The results of fluidity are demonstrated in **Figure** 7. it can be seen that increasing PET content tends to increase the flow-out time and hence causes a reduction in fluidity. However, no

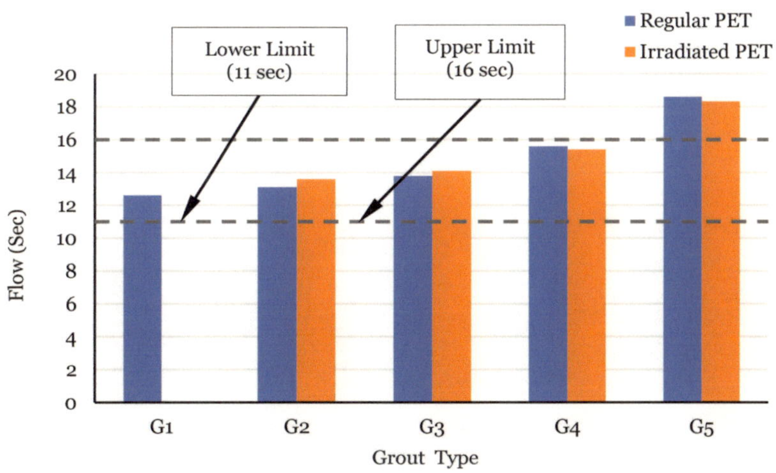

Figure 7.
Fluidity of cement grouts containing PET and FA.

significant change occurs due to the irradiation of the PET. Moreover, the fluidity of all grouts is within the range of 11 sec to 16 sec except grout containing 10% PET (regular and irradiated). The reduction in the fluidity of grout containing PET is due to the relatively large particle size of PET as compared to cement which restricts the free movement of a particle in the grouts.

Similar behavior is also noticed in grouts containing PET and silica fume as illustrated in **Figure 8**. The fluidity decreased with increasing PET content in the presence of silica fume as compared to controlled grouts. However, the fluidity of grouts containing silica fume is quite lower than that of grouts containing fly ash at the same PET contents. In this case, the fluidity of grouts containing 7.5% PET and 10% PET does not lie within the required range and hence cannot be recommended for SFP surfaces.

7.2 Strength of grouts containing irradiated PET

The compressive strength of grouts was determined by fabricating cube specimens of dimension 50 mm *50 mm *50 mm. After demolding, the specimens were cured in water and tested for compression test at 7-days and 28-days curing age. The results are illustrated in **Figures 9–12**.

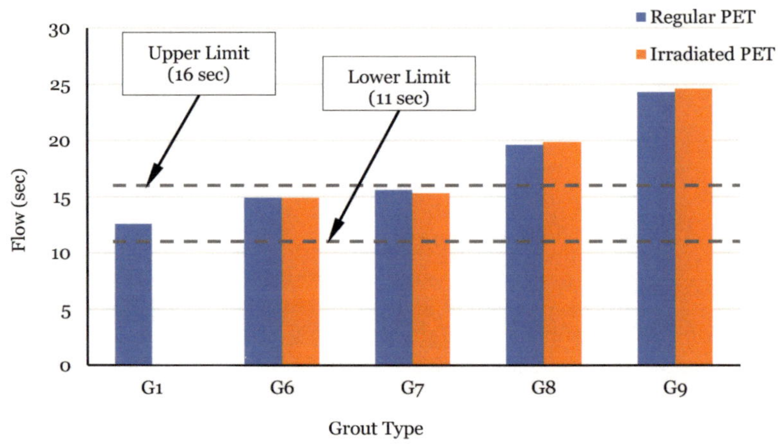

Figure 8.
Fluidity of cement grouts containing PET and SF.

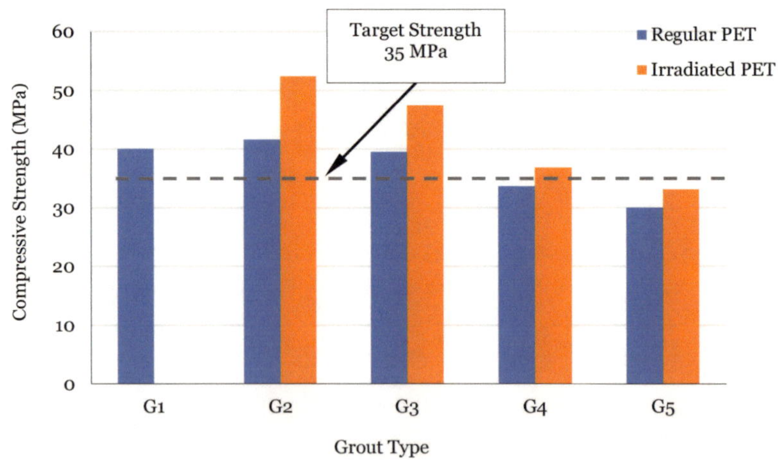

Figure 9.
Compressive strength (7-days) of grouts containing PET and FA.

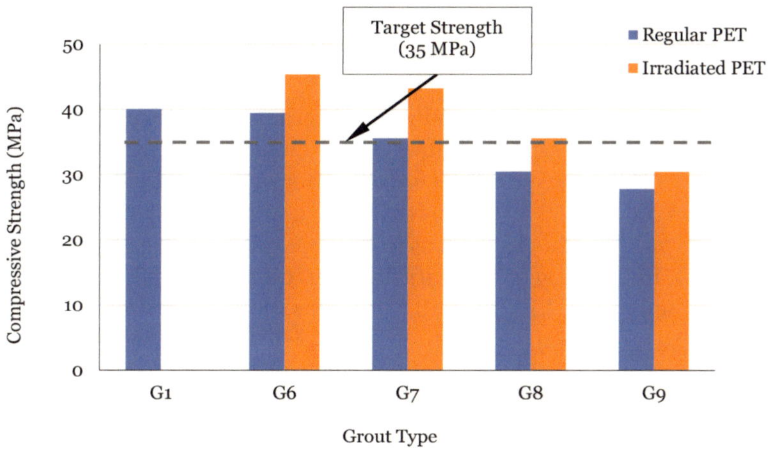

Figure 10.
Compressive strength (7-days) of grouts containing PET and SF.

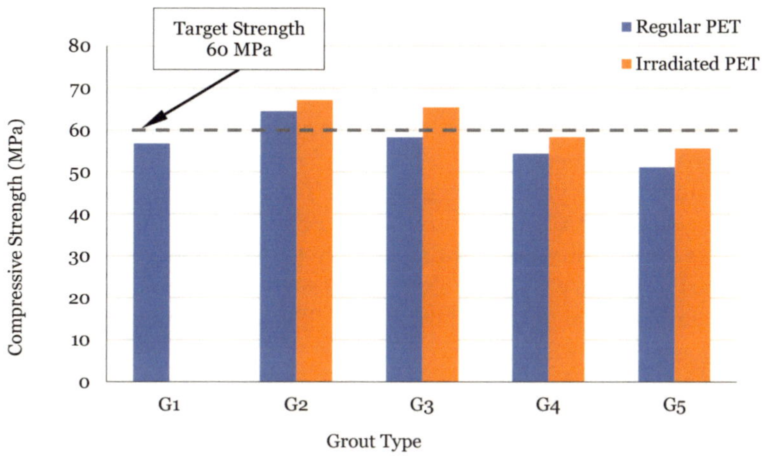

Figure 11.
Compressive strength (28-days) of grouts containing PET and FA.

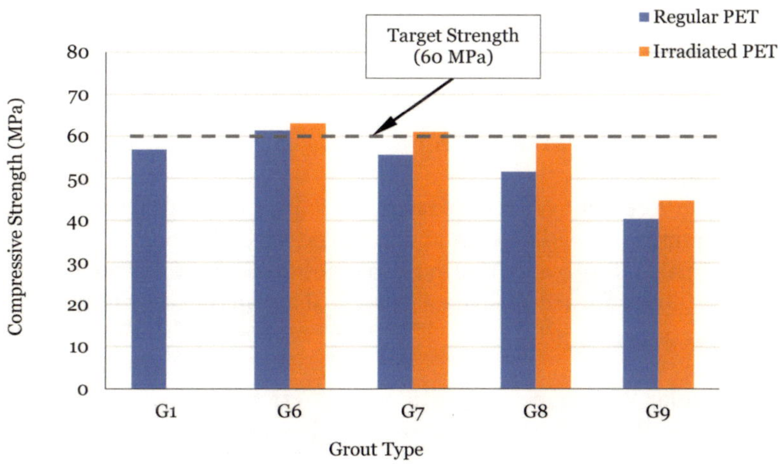

Figure 12.
Compressive strength (28-days) of grouts containing PET and SF.

It is a well-known fact that replacing cement with PET causes a significant reduction in strength properties. On the other hand, compressive strength is improved while using 10% FA as a cement replacement and in combination with a superplasticizer. It can be seen from **Figure 9**, grout containing 2.5% regular PET and 10% FA (G2) has slightly higher 7-days compressive strength compared to controlled grout (G1). Despite 10% FA, all other grouts containing regular PET (i.e 5%PET, 7.5%PET, and 10%PET) have 7-days compressive strength lower than that of control grout. It is due to the loss in compressive strength while replacing cement with PET. Interestingly, a significant amount of strength has been recovered while using irradiated PET instead of regular PET as show in G2 to G5. The strength recovery is higher at a lower percentage of PET (i.e 2.5% and 5%) and lower at higher percentages. The purpose of this study was to recycle the maximum percentage of waste PET while achieving the target 7-days compressive strength of 35 MPa. Hence, it can be seen that the cement grout containing 7.5% regular PET does not achieve the target strength, however, with the irradiation the same dosage of PET replacement reached the target strength. Similar behavior is noticed in the grouts containing silica fume and PET as shown in **Figure 10**. However, the 7-days compressive strength of all grouts containing regular PET (G6 to G9) is lower than controlled grout (G1).

Moreover, 28-days compressive strength gives a more clear picture to decide the final maximum percentage of PET that can replace cement in the grouts. **Figure 11** depicts that grout with 2.5% PET (both regular and irradiated) achieved the target strength of 60 MPa. However, grout containing 5% of regular PET does not reach the target line. Interestingly, 5% of irradiated PET recovered the strength loss and reached the target strength. Furthermore, 7.5% PET and 10% PET (both regular and irradiated) have lower compressive strength than the target value and hence cannot be recommended. The very similar behavior is also witnessed in the grouts containing silica fume in addition to regular and irradiated PET as shown in **Figure 12**. Grout containing 5% irradiated PET just achieved the target strength of 60 MPa while the strength of 5% regular PET is quite lower than the target line. Moreover, the overall strength of grouts containing FA is higher than that of grouts containing SF.

8. Conclusion

In the current study, cement in the grouts was replaced with PET (regular and irradiated), fly ash and silica fume and were evaluated for flowability and strength properties. Following conclusions are summarized from this study.

The air voids content in the range of 25–35% is generally required in mix designing the open-graded asphalt mixtures for semi-flexible pavement surfaces. These voids are required to ensure the interconnectivity of vids in the mixture to allow grout infiltration.

Highly flowable cement grouts are required to properly fill the voids in the open graded asphalt mix. Utilization of superplasticizer not only reduce the water-cement ratio but also improve strength properties up to some extent. Similarly, the pozzolanic reaction of fly ash and silica fume is also enhanced by using superplasticizer and help in improving 7-days and 28-days compressive strength as compared to controlled grout.

Utilizing regular PET causes a significant reduction in the compressive strength of grout. However, the irradiation process of PET (by exposing to gamma rays) recovers some strength which was lost due to regular PET. Similarly, the compressive strength was also improved by additional replacement of cement with fly ash and silica fume.

In achieving the target fluidity of 11–16 sec and 28-days compressive strength of 60 MPa, a maximum of 2.5% regular PET can be used as a cement replacement in addition to fly ash and silica fume. Interestingly, the same range of fluidity and target strength was achieved with 5% irradiated PET. Hence, the irradiation of PET can be used as an effective way to recycle a relatively large amount of plastics in the construction industry to achieve sustainability.

Finally, 0.35 w/c ratio, 1% superplasticizer, 5% irradiated PET in addition to 10% fly ash or 5% silica fume are recommended as a composition of cement grouts that can be used in designing semi-flexible pavement surfaces with an aim of sustainability and reducing greenhouse gas emission.

Author details

Muhammad Imran Khan[1*], Muslich Hartadi Sutanto[1], Madzlan Bin Napiah[1] and Salah E. Zoorob[2]

1 Department of Civil and Environmental Engineering, Universiti Teknologi PETRONAS, Malaysia

2 Construction and Building Materials Program, Kuwait Institute for Scientific Research, Kuwait

*Address all correspondence to: muhammad_17007177@utp.edu.my

References

[1] Zarei, S., et al., Experimental analysis of semi-flexible pavement by using an appropriate cement asphalt emulsion paste. Construction and Building Materials, 2020. **230**: p. 116994.

[2] Cai, J., et al., Comprehensive service properties evaluation of composite grouting materials with high-performance cement paste for semi-flexible pavement. Construction and Building Materials, 2017. **153**: p. 544-556.

[3] Abedini, M., et al., Low-temperature adhesion performance of polymer-modified Bitumen emulsion in chip seals using different SBR latexes. Petroleum Science and Technology, 2017. **35**(1): p. 59-65.

[4] Moghaddam, T.B., M.R. Karim, and M. Abdelaziz, A review on fatigue and rutting performance of asphalt mixes. Scientific Research and Essays, 2011. **6**(4): p. 670-682.

[5] Xu, T. and X. Huang, Investigation into causes of in-place rutting in asphalt pavement. Construction and Building Materials, 2012. **28**(1): p. 525-530.

[6] Hou, S., T. Xu, and K. Huang, Investigation into engineering properties and strength mechanism of grouted macadam composite materials. International Journal of Pavement Engineering, 2016. **17**(10): p. 878-886.

[7] Ling, T.-Q., et al., The Application of Semi-Flexible Pavement on Heavy Traffic Roads. International Journal of Pavement Research and Technology, 2009. **2**(5).

[8] Oliveira, J.R.M., N.H. Thom, and S.E. Zoorob, Design of Pavements Incorporating Grouted Macadams. Journal of Transportation Engineering, 2008. **134**(1): p. 7-14.

[9] Disfani, M.M., et al., Performance evaluation of semi-flexible permeable pavements under cyclic loads. International Journal of Pavement Engineering, 2020. **21**(3): p. 336-346.

[10] Pratelli, C., et al., Preliminary In-Situ Evaluation of an Innovative, Semi-Flexible Pavement Wearing Course Mixture Using Fast Falling Weight Deflectometer. Materials (Basel, Switzerland), 2018. **11**(4): p. 611.

[11] Tran, T.N., et al. Semi-flexible material: The sustainable alternative for the use of conventional road materials in heavy-duty pavement. in Congrès International de Géotechnique–Ouvrages–Structures. 2017. Springer.

[12] LING, T.-q., et al., Research on performance of water-retention and temperature-fall semi-flexible pavement material. China Journal of Highway and Transport, 2010. **2**.

[13] Al-Qadi, I.L., H. Gouru, and R. Weyers, Asphalt Portland cement concrete composite: laboratory evaluation. Journal of transportation engineering, 1994. **120**(1): p. 94-108.

[14] Pei, J., et al., Design and performance validation of high-performance cement paste as a grouting material for semi-flexible pavement. Construction and Building Materials, 2016. **126**: p. 206-217.

[15] Afonso, M.L., et al., Development of a semi-flexible heavy duty pavement surfacing incorporating recycled and waste aggregates – Preliminary study. Construction and Building Materials, 2016. **102**: p. 155-161.

[16] Yang, B. and X. Weng, The influence on the durability of semi-flexible airport pavement materials to cyclic wheel load test. Construction and Building Materials, 2015. **98**: p. 171-175.

[17] Hirato, T., M. Murayama, and H. Sasaki, Development of high stability hot mix asphalt concrete with hybrid binder. Journal of Traffic and Transportation Engineering (English Edition), 2014. **1**(6): p. 424-431.

[18] Hassani, A., M. Taghipoor, and M.M. Karimi, A state of the art of semi-flexible pavements: Introduction, design, and performance. Construction and Building Materials, 2020. **253**: p. 119196.

[19] Cai, X., et al., Interlocking property evaluation of dual skeleton in semi-flexible pavement material by micromechanical model and X-ray computed tomography. Construction and Building Materials, 2020. **254**: p. 118934.

[20] Oliveira, J.R., Grouted macadam: material characterisation for pavement design. 2006, University of Nottingham: Nottingham.

[21] Corradini, A., et al., Improved understanding of grouted mixture fatigue behavior under indirect tensile test configuration. Construction and Building Materials, 2017. **155**: p. 910-918.

[22] An, S., et al., Laboratory and Field Evaluation of a Novel Cement Grout Asphalt Composite. Journal of Materials in Civil Engineering, 2018. **30**(8): p. 04018179.

[23] Gong, M., et al., Evaluation on the cracking resistance of semi-flexible pavement mixture by laboratory research and field validation. Construction and Building Materials, 2019. **207**: p. 387-395.

[24] Hassan, K., A. Setyawan, and S. Zoorob. Effect of cementitious grouts on the properties of semi-flexible bituminous pavements. in Proceedings of the 4th European Symposium on Performance of Bituminous and Hydraulic Materials In Pavements, Bitmat 4, 11-12 April 2002, Nottingham, UK. 2002. Routledge.

[25] Cihackova, P., et al., Performance characteristics of the open-graded asphalt concrete filled with a special cement grout. Baltic Journal of Road and Bridge Engineering, 2015. **10**(4): p. 316-324.

[26] Saboo, N., et al., Development of hierarchical ranking strategy for the asphalt skeleton in semi-flexible pavement. Construction and Building Materials, 2019. **201**: p. 149-158.

[27] Saboo, N., et al., Optimal proportioning of grout constituents using mathematical programming for semi flexible pavement. International Journal of Pavement Research and Technology, 2019. **12**(3): p. 297-306.

[28] Van de Ven, M. and A. Molenaar, Mechanical Characterization of Combi-layer. Asphalt paving technology, 2004. **73**: p. 1-22.

[29] Anderton, G.L., Engineering properties of resin modified pavement (RMP) for mechanistic design. 2000: U.S. Army Corps of Engineers, Engineer Research and Development Center.

[30] Setyawan, A., Development of semi-flexible heavy-duty pavements. 2006, University of Leeds.

[31] Ahlrich, R.C. and G.L. Anderton, Construction and Evaluation of Resin Modified Pavement. 1991, Army Engineer Waterways Experiment Station Vicksburg MS Geotechnical Lab: Army Engineer Waterways Experiment Station Vicksburg MS Geotechnical Lab.

[32] British Board of Agrement (BBA), Hardicrete Heavy Duty Surfacing. 1994, British Board of Agrement (BBA), Agrement Certificate No 88/1969. Watford.

[33] British Board of Agrement (BBA), Worthycim Heavy Duty Paving. 1996, British Board of Agrement (BBA), Agrement Certificate 87/1900. Watford.

[34] ApS, C. A Jointless Wearing Course with High Strength for High Loads. 2018 [cited 2018 December]; Available from: http://www.confalt.com/confaltr.html.

[35] Densit-a/s, "Densiphalt Handbook". 2000, Aalborg,.

[36] Ahlrich, R.C. and G.L. Anderton. Resin Modified Pavement in Airfield Applications. in Airport Pavement Innovations. Theory to Practice. Proceedings of Conference Held Sept 8-10, 1993, Vicksburg, Mississippi, USA. 1993.

[37] Ricci, R., Laboratory study of grouted macadams impregnated with mine waste geopolymeric binder. 2012, University Di Bologna.

[38] Anderton, G.L., Engineering properties of resin modified pavement (RMP) for mechanistic design. 2000: Engineer Research and Development Center Vicksburg MS Geotechnical Lab.

[39] Vavrik, W.R., et al., The bailey method of gradation evaluation: the influence of aggregate gradation and packing characteristics on voids in the mineral aggregate (with discussion). Journal of the Association of Asphalt Paving Technologists, 2001. **70**.

[40] AASHTO T 305, Standard method of test for determination of draindown characteristics in uncompacted asphalt mixtures. 2005, American Association of State Highway and Transportation Officials.

[41] ASTM D6390, Standard Test Method for Determination of Draindown Characteristics in Uncompacted Asphalt Mixtures, in ASTM International, West Conshohocken, PA. 2017, ASTM International, West Conshohocken, PA: ASTM International, West Conshohocken, PA.

[42] Hou, S., T. Xu, and K. Huang, Aggregate Gradation Influence on Grouting Results and Mix Design of Asphalt Mixture Skeleton for Semi-Flexible Pavement. Journal of Testing and Evaluation, 2017. **45**(2): p. 591-600.

[43] REAM, Road Engineering Association of Malaysia, "Speciication of Semi-Rigid Wearing Course". 2007, Road Engineering Association of Malaysia (REAM).

[44] P.W.D, Public Works Department "Description of Workmanship and Materials (Semi-Rigid Pavement), Section E" City Hall Kuala Lumpur, Kuala Lumpur, Malaysia, 2003. 2003.

[45] Anagnostopoulos, C.A., Effect of different superplasticisers on the physical and mechanical properties of cement grouts. Construction and Building Materials, 2014. **50**: p. 162-168.

[46] Imran Khan, M., et al., Optimization of Cementitious Grouts for Semi-Flexible Pavement Surfaces Using Response Surface Methodology. IOP Conference Series: Earth and Environmental Science, 2020. **498**: p. 012004.

[47] Zhang, J., et al., Formulation and performance comparison of grouting materials for semi-flexible pavement. Construction and Building Materials, 2016. **115**: p. 582-592.

[48] Koting, S., et al., Effects of Using Silica Fume and Polycarboxylate-Type Superplasticizer on Physical Properties of Cementitious Grout Mixtures for Semiflexible Pavement Surfacing. The Scientific World Journal, 2014. **2014**: p. 7.

[49] Zhong, K., et al., Interfacial and mechanical performance of grouted open-graded asphalt concrete with latex modified cement mortar. Construction and Building Materials, 2020. **234**: p. 117394.

[50] Xu, Y., et al., High-Performance Semi-Flexible Pavement Coating Material with the Microscopic Interface

Optimization. Coatings, 2020. **10**(3): p. 268.

[51] Luo, S., et al., Open-graded asphalt concrete grouted by latex modified cement mortar. Road Materials and Pavement Design, 2018. **21**(1): p. 1-17.

[52] Wang, D., et al., Impact analysis of Carboxyl Latex on the performance of semi-flexible pavement using warm-mix technology. Construction and Building Materials, 2018. **179**: p. 566-575.

[53] Zhang, W., et al., Performance Characterization of Semi-Flexible Composite Mixture. Materials, 2020. **13**(2): p. 342.

[54] Cai, X., et al., Identification of microstructural characteristics in semi-flexible pavement material using micromechanics and nano-techniques. Construction and Building Materials, 2020. **246**: p. 118426.

[55] Zhang, H., et al., Study on the mechanical performance and application of the composite cement–asphalt mixture. International Journal of Pavement Engineering, 2019. **20**(1): p. 44-52.